ANDERSON'S WAY

ANDERSON'S WAY

THE STORY OF AN ENTREPRENEUR

Joseph L. Massie

WCB Brown & Benchmark
PUBLISHERS

Madison, Wisconsin • Dubuque, Iowa

Book Team

Editor *Paul Tavener*
Production Editor *Jane C. Morgan*
Visuals/Design Developmental Consultant *Marilyn A. Phelps*
Visuals/Design Freelance Specialist *Mary L. Christianson*
Production Manager *Beth Kundert*

WCB Brown & Benchmark

A Division of Wm. C. Brown Communications, Inc.

Executive Vice President/General Manager *Thomas E. Doran*
Vice President/Editor in Chief *Edgar J. Laube*
Vice President/Marketing and Sales Systems *Eric Ziegler*
Director of Production *Vickie Putman Caughron*
Director of Custom and Electronic Publishing *Chris Rogers*

Wm. C. Brown Communications, Inc.

President and Chief Executive Officer *G. Franklin Lewis*
Senior Vice President, Operations *James H. Higby*
Corporate Senior Vice President and Chief Financial Officer *Robert Chesterman*
Corporate Senior Vice President and President of Manufacturing *Roger Meyer*

Dust jacket, cover, and interior designed by Terri W. Ellerbach

Library of Congress Catalog Card Number: 93–74977

ISBN 0–697–24156–4

Printed in the United States of America by Wm. C. Brown Communications, Inc.,
2460 Kerper Boulevard, Dubuque, IA 52001

10   9   8   7   6   5   4   3   2   1

CONTENTS

FREE ENTERPRISE, ENTREPRENEURSHIP, RALPH ANDERSON, AND BELCAN

ANDERSON'S PHILOSOPHY

EDUCATION AND PERSONAL DEVELOPMENT, 1923–1957
ANDERSON'S LIFE-LONG KENTUCKY IDENTIFICATION

BELCAN: TEMPORARY ENGINEERING SERVICES,
1958–1975

FULL SERVICE ENGINEERING—STAGE OF RAPID GROWTH, 1976–1985

FINANCIAL CRISIS AND CONSOLIDATION, 1986–1989

INTEGRATED STRATEGIC PLAN

NOT ONLY THE COMPANY ONE KEEPS, BUT THE PEOPLE THE COMPANY KEEPS

FIFTY-ONE PERCENT LAUGHTER, THE REST WISDOM AND LOVE

LIST OF EXHIBITS

ALPHABETICAL LISTING OF PROFILES

S Since the building of a company involves attracting appropriate personnel, we offer the profiles of 31 of these people as they appear in the chronology at the end of the chapter in which he or she joins the company. This placement has several advantages: 1. The description of the person helps round out the role that he took at the time. 2. Facts in the profile provide further insight into the facts in the chronology. 3. The facts in the profile provide a solid background for the discussion in Chapter Eight where we analyze the method in which Anderson chose people and they chose Belcan. This alphabetical listing offers quick access to the stories of these key individuals.

FOREWORD

I It is an honor to write the foreword to this book. It is a great book, about a great man, who achieved great things. *Anderson's Way.* What a way to go! Positive. Hard working. Honest. Creative. Productive. Generous. Daring. Optimistic. These are the road signs along Anderson's Way.

As you read this book you will see how these qualities of character all come together in the life of my dear friend, Ralph Anderson.

He dreamed his dream and then dared to put his dream to work, following through with focus until he achieved and then moving on to the next challenge. Always dreaming, always learning, always working, always achieving. This is the formula for Anderson's success.

As this book unfolds you will discover the making of a true role model. We will all learn much from reading about his life. We all admire the entrepreneur. Here is a living example of that lifestyle. Ralph Anderson brings people, ideas, science, and marketing all together. He practices one of the great lessons in life—that you can't do it alone. I have often said that the success I have enjoyed is because the right people were in the right place at the right time to help me. This is also the story of Anderson's Way. Learning how to find those people, attract them to you and your mission, and hold their loyalty and support—that is the key. You will see it unfold before your eyes in this book. It is truly a "help-help" equation. You can succeed if you understand and practice this principle.

I know Ralph Anderson well. He is a member of my International Board of the Crystal Cathedral Ministries. I am proud to have him by my side to lend his wisdom, work ethic, and support. Most of all he is my friend. Ralph and his dear wife Ruth are special together. They are an amazing team. Their happiness comes through in the way they look and act. I call them my "happy face friends." In these people we discover those rare qualities that balance simplicity, joy, brilliance, firmness, faith, hope, and love.

This book will reveal these qualities. Read every word carefully. Just as Anderson's Way is always searching for a better way, so will readers be led into new ideas and practices for the personal enrichment of their lives. You will learn a lot from this book! I have!

Dr. Robert H. Schuller
Crystal Cathedral
Garden Grove, California

PREFACE

T This book was written because it was FUN. My previous books were for academic recognition, published research, and royalties from textbooks, but after my retirement in 1986, I wanted a change. After traveling with my wife to most countries of interest to us, I returned to my early interest in entrepreneurship. I started by consulting with Belcan and recording Anderson's story as an oral history project for King Library at the University of Kentucky. I was then asked by employees of Belcan to write the story of Ralph Anderson and that of Belcan.

I made an oral agreement with Anderson. Using a tape recorder, I proceeded without deadlines or publishers' contracts. It became obvious that we were enjoying talking! Furthermore, as my visits with Anderson expanded to include the people of Belcan, the focus changed from Anderson himself to include the many interesting people who had been attracted to him.

Upgrading from a manual typewriter in my past writing to a word processor for this book, I still found that Belcan's experts could keep up with the computer technology better than I could. I became dependent on Kathleen McClain from the Communications Department for editorial assistance and on-location research.

Interviewing and writing this short book became so much fun for me and for Ralph that some of our friends wondered whether it would ever be finished. As you will see in the following chapters, our book did get finished and Ralph continues to have fun in whatever he does. I hope that readers not only get ideas about entrepreneurship in the following pages, but they may consider the possibility of writing for the fun of it! I recommend doing your own thing both in business and in retirement!

Joseph L. Massie

ACKNOWLEDGMENTS

Since a large part of this book is about people, particularly their verbal comments, the writer is indebted to all those who gave willingly of their time in taped interviews. Of course, the one devoting the most time was Ralph Anderson. Beginning with the initial oral history interviews for the University of Kentucky, Anderson offered his complete cooperation. He facilitated contacts with childhood friends, business associates, and competitors.

All officers and top level managers of Belcan were interviewed in some depth. Key staff personnel such as Jane York and Barbara Schmidt were most supportive. From these interviews developed the basis for the profiles. Also, I want to mention the help given to me by Susan McFarland and Terri Faul.

Anderson's early childhood friends were eager to contribute: Frances Board Keightley, a friend from the first grade, John Harris, a friend from the seventh grade, Charles Noel and John Sullivan, friends from Harrodsburg High School.

Karl Schakel, Fort Collins, Colorado, gave key information about Anderson just before Belcan was formed. Jack Hope, Hillsboro, Ohio, covered forty years of their association. Phil Bright and K. O. Johnson were helpful in telling about Anderson as a practicing engineer.

Interviews with Ruth and Candace Anderson McCaw gave the family viewpoint. Ruth Anderson and Nita Yoder (deceased) provided details of the first ten years of the Belcan office activities.

Views of a competitor, Gus G. Perdikakis, of a consultant, James R. HusVar, and a former personal secretary, Mrs. Mari Otto, enriched the details of Anderson's approach.

Recognition should be given to several artists: Edith L. Clifton, the painter of Anderson Circle Farm in winter, James Werline, the painter of Walnut Hall, and David Mueller, the painter of the portrait of Anderson.

Tom Philpot drew the J. J. Suarez profile caricature. Beth Gilbert was chief photographer, including the portrait, the modern photos on the dust cover of the book, the photos of the family, and the photos of Anderson Circle Farm. David Monhollen carved the eagle also used on the dust cover.

Special thanks and recognition go to David K. Blythe, Professor Emeritus of Civil Engineering, University of Kentucky, who taught Anderson in the engineering program in the 1940s. Dave introduced the author to Anderson in the 1980s and recommended him to Anderson for both the oral history research and the development of this book. It is true that this book literally would not have been written if Blythe had not promoted the idea to both the subject and to the author.

A second person to whom special thanks and recognition is due is Ms. Kathleen (Katie) McClain, Writer/Producer, Belcan Video Productions, a Division of Belcan Engineering Group, Inc. Katie filled the crucial role of editor within the company and provided professional service during the manuscript stage of the book. Katie wore many hats including agile communicator, creative advisor, and friend.

My chief consultant in all my writings, and especially in the final stages of this book has my thanks, highest respect and deepest love—my wife for over 50 years, Christine (Chris) Massie.

Free Enterprise, Entrepreneurship, Ralph Anderson, and Belcan

"Entrepreneurs create jobs and hire managers."
J. L. Massie

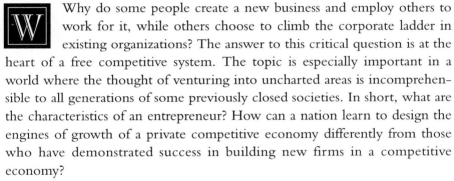

Why do some people create a new business and employ others to work for it, while others choose to climb the corporate ladder in existing organizations? The answer to this critical question is at the heart of a free competitive system. The topic is especially important in a world where the thought of venturing into uncharted areas is incomprehensible to all generations of some previously closed societies. In short, what are the characteristics of an entrepreneur? How can a nation learn to design the engines of growth of a private competitive economy differently from those who have demonstrated success in building new firms in a competitive economy?

The thesis of this book maintains that some of the answers to these questions can be found by studying the characteristics of those persons who have started new companies, and how these companies developed into vibrant and successful organizations that gave employment to a large number of people. Each case tends to have unique elements; few facts are the same. The diversity of the approaches of individual entrepreneurs yields exciting and challenging stories.*

This book is one such story of an individual and the company that he created. Ralph Anderson is the person and Belcan Corporation is the company. The following chapters, then, are in a sense a biography and a

---

*See Joseph J. Jacobs, THE ANATOMY OF AN ENTREPRENEUR, FAMILY,
    CULTURE, AND ETHICS. 1991, San Francisco, California, ICS Press, Also, see over
    1,000 publications of the Newcomen Society in North America: Princeton University Press.

company history. The goal of the author is to provide a story of both the individual and the company for others to understand as they strive toward fulfilling their aspirations for their life's work.

Definition and Conceptual Orientation

First, let us clarify our conceptual foundation and provide a definition of the critical term. An entrepreneur can be defined as an organizer of an economic venture, especially one who organizes, owns, manages, and assumes the risk of a business. The critical elements of this definition are:

- an economic venture
- ownership of a business
- willingness to assume risks

An entrepreneur must perform the three elements listed above; he or she can hire others to manage and do other things. In fact, the characteristics of a good entrepreneur may conflict with the characteristics of a good manager, but the chief point is that he hires others, increases employment, and improves competition. In short, an **entrepreneur** creates a business that serves as the engine of growth for an economy. A **manager** organizes, plans and controls a going concern.

Most of the facts in the following chapters concern Anderson and Belcan, but we delve further, digging into the implications for an economic system and into the reasons why Anderson is motivated to serve the national economy through self-interest. Hopefully, it might provide ideas for others who wish to create companies of their own.

Besides offering an "overview" in this first chapter, we also want to provide an "inner view" of this entrepreneur and of the corporation he founded: what characteristics did Ralph Anderson possess that supported his building of the Belcan Corporation? What were Anderson's motivations? What personal traits did associates observe that influenced his actions? Finally, we summarize the birth and growth of Belcan Corporation from 1958 to 1993, in Appendix B as a guide for following the details in succeeding chapters.

Characteristics of an Entrepreneur

In one comprehensive study of entrepreneurs, over forty characteristics were identified. For our purposes, it is preferable to separate the ten most important and to distinguish entrepreneurship from closely related terms such as management and leadership. The ten characteristics of an entrepreneur are:

1. Comfortable with and even eager to assume risks.
2. Optimistic toward future.
3. Strives for independence.
4. Has the ability to make decisions quickly and remains flexible.
5. Eager to learn from mistakes and to initiate change.
6. Expects honesty and integrity.
7. Trusts others, for the most part, and enjoys working with people.
8. Creative and resourceful.
9. Possesses high energy and determination. Immediately identifiable as a hard worker.
10. Has ability to influence and get along with all types of people.

Notice some of the characteristics that were *not* included in the above list. Some of the following may be helpful but they are not required. In fact, some of the following qualities *conflict* with the above characteristics.

- Good management skills, including organizing, planning, staffing and controlling
- Values stability, bureaucracy, routine, going by the rules
- Interest in details and precision
- Sticks to a single direction
- Consistency
- High intellect
- Structured thought
- Theoretical thought

Entrepreneurs Differ from Managers

A basic thesis of this study is that entrepreneurship is a concept separate from management. Furthermore, some functions of management may not be conducive to an entrepreneurial effort. The extreme creativity of a successful entrepreneur may be inconsistent with planning and control, two important functions of a manager. If this is true, then a successful entrepreneur may need to hire a manager (or marry one!) to perform managerial functions. An entrepreneur may develop managerial functions or managerial abilities as learned from experience. In the case of Ralph Anderson, entrepreneurial characteristics are clearly evident in his personality. Ruth Anderson is a classic example of an organizer, a true manager, easily controlling activities. As our profile of Ralph Anderson develops, we shall observe that while remaining an entrepreneur, he directs more and more energy to managerial thought.

In 1991, as research for this book began, free enterprise and its engine of development, entrepreneurship, had clearly surpassed the ideas of Marx and Communism. During the research and initial writing of this book, world conditions set the stage for the topic to be in vogue in discussions of economics and political systems. The subject had been developed in some depth after World War II.

General Doriot, a unique professor of the Harvard Business School, formed a corporation, the American Research and Development Corporation, specifically to provide seed money for the formation of new enterprises. Route 128 around Boston became the locus of some new firms and later, Silicon Valley, south of Palo Alto, developed in the prime of the computer industry.

Those with experience in both hotbeds of capitalism began a renewed effort to provide research and books for this exploding field. One of these books, written by C. Gordon Bell with John E. McNamara, *High Tech Ventures: The Guide for Entrepreneurial Success,* was published by Addison Wesley in 1991. This book presents a framework for the formation of new firms. In 1958, Ralph Anderson had no such guide, and probably would not have used it if it had existed, because Anderson fits no single profile. The book is an excellent standard for approaching the formation of a new business. It provides us with a model around which we can study the characteristics of Ralph Anderson.

Stages of Start Up, Considerations for Excellence in Entrepreneurship

The essence of Bell's approach consisted of five stages of start up, twelve dimensions of consideration, and key generalizations for better understanding of entrepreneurship. The five stages of start up are:

- **Stage I**—Concept
- **Stage II**—Seed-the concept is defined and a detailed plan formed
- **Stage III**—Product Development-concept tested by users
- **Stage IV**—Market Development-product sold and company becomes profitable, proving its viability
- **Stage V**—Steady Stage-firm achieves liquidity, goes public, merges with another, or continues in private hands

Bell summarized some of the characteristics of the CEO of a new firm: leader, coach, manager and standards setter.*

Silver (1985) believes that the typical entrepreneur is a happy, creative, insightful, guilt-laden, 27–33 year old, who is a good communicator, comes from a middle-class home with an absent father, had a deprived childhood, is married or divorced, and can focus intensively for long periods of time. Both White (1977) and Silver believe that being wealthy is a significant handicap to an entrepreneur.

Bell lists key personal qualities exhibited by effective CEOs:

- Intelligent and energy
- Integrity, quality and working habits
- Environment
- Openness
- Good training and good role models are common attributes
- Team-building skills and the ability to delegate
- Ego and humility

With this summary of the subject, it is easier for us to compare and contrast Anderson's case. We will see very quickly that Anderson differs significantly in several ways. Belcan's product is customer service and thus, Anderson's concept of product was people focused.

Hypothesis Regarding the Entrepreneurial Personality

1. Is the field of engineering conducive to the ten characteristics that were listed prior to the more structured concepts of management? Is an entrepreneur-engineer mutually contradictory in some aspect? Without developing a theoretical model and belaboring the point, wouldn't it be interesting to study an individual who is both an engineer and an entrepreneur? The author answered this last question clearly in the affirmative; it is the reason that research and writing this book was enjoyable.

2. Are there similarities between a compulsive gambler and a compulsive entrepreneur? Both love to take risks and may get "hooked." The gambler likes risks so much he or she will *create a risk* so that he or she can bet on something; for example, which bug will reach the wall first?

*C. Gordon Bell, *High Tech Ventures: The Guide for Entrepreneurial Success*, 1991, Addison-Wesley Publishing Company, pp. 5-9.

The gambler produces nothing but his own pleasure. The entrepreneur may love risks just as much but he strives to "win" by creating a firm and by giving people employment. The gambler may be a leech on society, whereas the entrepreneur is the engine for growth in a free enterprise system.

3. Are entrepreneurs born or can they be made? We shall not try to analyze the difference between the effect of environment and heredity on the growth of entrepreneurs. A recent study of genes does raise questions as to how individuals might be programmed from birth for certain diseases and activities just as computers can be programmed for their functions.

We broach these questions only to provide background for deeper thought. The following chapters describe Anderson's characteristics that resulted in the creation of Belcan without attempting to separate those with which he was born from those acquired from experience. We outline briefly the genealogy of Anderson and the cultural environment into which he was born. We then discuss the evolution of his ideas as he first tried to become an "organization man." We observe that he soon learned that he had a hard time fitting into existing corporations.

Motivation of Ralph Anderson

Throughout the interviews and research, the author has been fascinated by questions such as, "What makes Ralph run?" Anderson has been frequently queried, "Why do you strive for another million dollars. Haven't you got enough?" At over 65 years of age, why not get around to relaxing?" "Why not play golf, travel to interesting places, write, or enjoy any of the many other opportunities available to a "retired" person?"

Certainly, the answers to these questions have been given in many scholarly books and novels. For example,

- Search for yet more power over others!
- Make more money—(a la' King Midas)
- Completion of a lifetime objective
- "Do good"
- Build security for descendants

When the author asked some of the above questions while chatting with Ralph and Ruth at the Mercer County Farm, Ralph paused, thought about it, and answered, "I do it for the fun of it." Ruth agreed that his answer was about as accurate as any she could imagine. Still, the answer was so simple that it could mislead and or be ignored. Only much later in the research did the author analyze the complexity of this simple answer.

Anderson participated in the preparation of the book not merely to gain "immortality," recognition, or his name in print. He worked on the book for the same reason that he worked for another million dollars—it was his idea of fun!

In his youth, Ralph Anderson had little time to play. Though he was mechanically inclined, he never got any mechanical toys at Christmas because his family had little funds for such non-essentials. As an adult, Ralph enjoys acquiring things he could not afford when he was a boy. His acquisitions run the spectrum from companies to fascinating widgets. At one point during the interviews, the author tested the hypothesis that Ralph was compensating in later life for things he didn't have as a boy. Anderson agreed, and added some amazing support. For example, he said, "I was never able to get a bike. Several times I was close, but I never did."

Now, in his 70s, Ralph Anderson can own practically anything money can buy. So what does he play with? He owns a Rolls Royce that he uses to drive visitors around his farm and serve his community as Grand Marshall of the parade opening the Mercer County Fair. He drives a 12-cylinder Mercedes, equipped with all of the options, on a daily basis. On the Mercer County farm he keeps a four wheel drive car so that he can show visitors the rugged areas near historical Shawnee Springs located on the Anderson Circle farm. Also, he has collected horse drawn carriages for special occasions.

Anderson possesses many different types of clocks—one hangs in the lobby at Anderson Way and is cared for by Anderson, himself. He found a clock that can be set to the precise time by radio from the atomic clock in Colorado. He liked this idea so much that he obtained a similar one for each of his officers so that there would be no question as to the exact time for scheduling business. Another clock at Walnut Hall shows nighttime and daytime graphically for each season in the northern and southern hemispheres, each hour of the day. One can quickly determine sunrise and sunset for any point on the world map.

Anderson collects unusual mechanical things: bizarre farm equipment, a shoulder-operated 40 mm "canon" developed by the late Colonel George Chinn (the only purpose of which might be to break one's shoulder if it were fired), and two large "coffee tables" that are scale models of his 3,000 acre Mercer farm. A magnificent grand piano (converted to cassette-driven player piano) graces the foyer of Walnut Hall. Anderson bought two go-carts and two four-wheelers, (ATVs), for his grandchildren. The list continues. An observer might conclude that Anderson is enjoying his second childhood—but he never actually had a first one. Therefore, throughout adult life, he continues to cultivate his first childhood.

For the last decade, the Andersons have travelled from Cincinnati to Mercer County to be in "home territory" as they acquired a 3,000 acre farm and oversaw the renovation of an 8,000 square foot mansion. Friends visit them primarily at the farm, rather than at their home in Indian Hill. Employees of the company have the opportunity to see the progress on the farm on an annual basis at the Belcan pig roast in the fall. The land was originally purchased as a means of storing assets (forced savings) but as the site increased, the farm became a business with a professional manager.

The analogy of a game may appear superficial. But looking into Anderson's interest in athletics at the University of Kentucky, one might see that his life changed as a result of attempting to return to Lexington for a football game. One of the few things that can tempt Anderson away from his farm on the weekend is a Wildcat basketball game. His interest in helping the university financially is connected with his interest in athletic events. Anderson has had this interest in U.K. athletics since the 1950s. Anderson views life itself as a competitor—he loves to win! Coach Rupp once said, "Winning isn't everything—it is the only thing!" For Ralph Anderson, the idea is different; he values action, trying, doing one's best. Anderson believes the objective is in the striving, not necessarily in the winning or the ultimate fruits of striving.

Ralph and Ruth grew up during the depression of the 1930s in families of limited means. They didn't associate with wealthy people. After achieving success, Ralph has enjoyed returning to his home county armed with a new status which he did not possess as a young person.

Anderson grew up on Warwick Road, just outside of Harrodsburg, Kentucky. Anderson's Mercer County Farm is located on both sides of Warwick Road. A citizen of Mercer County, whether he be from the old, landowning class of the county or a small farmer, cannot escape the fact that a successful local boy had returned to the county.

Ralph and Ruth enjoy being together and doing similar things. As Ruth commented, "Ralph and I agreed that one dollar is not Ralph's and another is Ruth's. Ralph's money is mine and mine is his."

Ralph encourages Ruth to collect her own toys. Ruth had few dolls as a young girl so now, in her adult life, she has enjoyed collecting dolls and having a doll house. Her collection includes U.S. Presidents' wives, baby dolls and school-aged dolls. Ruth places her dolls in creative and unexpected spots around the Andersons' Cincinnati home and at Walnut Hall, perhaps seated in a miniature rocker by the fireplace or sleeping peacefully in a an old-fashioned pram in the corner of a guest room.

Ruth loves shopping for unusual things, searching for good buys. She once purchased a large wooden train for $100 which was used for Christmas decorations by a large department store in Cincinnati. Then Ruth relocated it to Walnut Hall, where the train would fit. Her comment was, "Don't you think that was a good buy?"

Anderson loves to talk about new ideas raised by others. He motivates the people around him to stretch their imagination and improve things. In the last ten years, with his increased age and status, many young people like to be around him and seek his advice. Anderson spends a fair amount of time giving advice (not limited to business advice) to friends who seek his council. Former employees call him long distance. Anderson never appears to be in a hurry and never cuts a person off abruptly. It is evident to those who seek his opinions that he is interested in their personal happiness. This same characteristic is a solid human relations skill used for reducing personnel turnover and increasing morale in the company. People who have come to truly know Ralph Anderson trust him, value his advice, and therefore, seek his council, even when their paths have led them away from Belcan.

Recently, Anderson has begun a collection of books and tapes by well-known people who concentrate on philosophical concepts, techniques for motivating people, and others who have recommendations for handling stress and the development of positive thinking. Anderson is not satisfied to purchase one tape or book but he buys several so that he can encourage others to seek basic answers to living. Thus, his help to friends is not only advice but also creative ideas of thinking. An admirable library of motivational audio cassettes, video programs and books are available at the Anderson Way Quality Resource Center (See Appendix C for a list of books and tapes) to be checked out by any employee.

It may be interesting to list a few of Anderson's personal traits and actions which have been observed by associates. These actions are inherent in the personal makeup of Anderson and have not been cultivated as a "recommended human resource prescription" by management gurus. They are important to understand because they represent a very important reason why many employees remain with Belcan and feel a personal commitment to its management. The word might be *morale,* but the personal relationship between Anderson and many others in the company resembles more of a friendship, at times a mentor, or fatherly advisor. The following are observations by a number of different associates who have been with Anderson for a period of years:

"Since he was a small boy, Ralph Anderson has been a hard worker . . . in school, in athletics, in business . . . in all activities."

"Anderson is always ready to take chances. Those who have contributed to his success feel little envy because they know that they could have profited from taking the risks; they didn't, Anderson did."

"Anderson is able to keep information in his head so that he can approximate data concerning revenue and expenses very close to the actual accounting monthly reports." (A bookkeeper commented that she was continually amazed how close he could come to the actual figures which she had recorded.)

"Anderson provides subordinates too much chance to take advantage of him; he does not maintain tight controls." (This characteristic tends to make individuals especially conscientious; they wish to treat him fairly since he is so trusting.)

"Ruth Anderson believes in her husband, supporting him in all his endeavors. When she suspects someone might be taking advantage of him, she makes herself especially aware of the detailed actions."

"Anderson's enthusiasm is contagious, as is his optimism. Chatting with Ralph is every bit as therapeutic as visiting a psychiatrist."

"Anderson listens with sincere interest to ideas and problems of others. He has the ability to place a person at ease, allowing the person to enjoy himself and think clearly."

An Entrepreneur's Role in Free Enterprise

The following pages include the story of a person who feels that he has been successful when measured by several different criteria.
His personal life has been satisfying by achieving perceived goals:

- remaining happily married for over 45 years, having a daughter and grandchildren
- remaining in contact with numerous friends over a long period
- maintaining healthy mental adjustments to life (i.e., peace with minimum stress)

- retaining free choice with minimum restrictions set by others
- maintaining physical health through mental control
- remaining active in work, regardless of age

His business life is summed up in the founding and growth of Belcan:

- starting with no personal or inherited wealth, i.e., self made!
- using an "average collection of genes" with little family prestige in a small town that values family status
- taking adverse business challenges in stride
- amassing equity in his own company through increasing gross revenue (See Figure 1.1)
- searching for what the client wants and providing that service

His contribution to society and to posterity as measured by:

- providing employment and challenges for Belcan personnel
- looking for new ideas for technology to contribute to society
- financially contributing to educational institutions, religious organizations, and community projects
- preserving natural resources and historical landmarks

Now that we have introduced the relationship of free enterprise and entrepreneurship to Ralph G. Anderson and Belcan Corporation, we proceed to both the biographical facts of Anderson (See Chapter Three and the highlights at back of book) and the historical facts of Belcan Corporation (See Chapters Four through Seven). Appendix B summarizes both.

As you read the early chapters, keep in mind that Anderson was 35 before he decided to incorporate Belcan. Even by age 40 he had acquired no equity. Belcan's future was so dim that Anderson could find little support from commercial banks. His sales ability started with his "education" of bankers and telling them that he would repay any loan he was granted. Later, he would stress the need for anyone to first "find out what the client wants" and then provide the client that service. A single word to describe his activities is spontaneity. When his "gut" tells him to move, he responds with confidence, enthusiasm and hard work.

Like most small firms without access to data banks, the most interesting facts about the early stages of the business depend on memory and on oral discussion using a tape recorder. The history that can be supported by a paper trail or written record is not much more than 15 years old. Even in

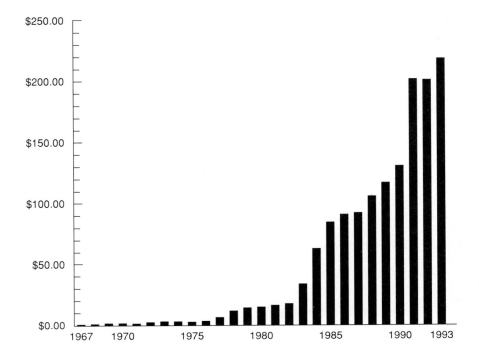

Dollar figures represent millions

Figure 1.1 Belcan Corporation Gross Revenue, 1958–1992.

the rapidly changing society of the last several decades this time could be viewed as relatively short. Furthermore, although the author became aware of Anderson and Belcan after his retirement, relating this short history to his personal observations of Belcan over six years makes him more confident of this interpretation. And when we relate this period to the fall of the communist threat and to the increased interest of the entire world about entrepreneurship, the following pages are not only interesting but timely.

ANDERSON'S PHILOSOPHY

"Think positively about your life and your work."
RALPH G. ANDERSON

Through the history of Belcan, Ralph Anderson developed a philosophy that formed the foundation for the company and its growth. We have seen that his personal characteristics explain the manner in which Belcan continues to grow. Certainly, luck played a role at times. But the story of Anderson and Belcan would be only a series of facts and anecdotes unless we focus on the principles that guide his actions. In this chapter we shall look into these principles.

There is a statement occasionally spotted on a bumper sticker which is the most concise way that we can summarize Ralph Anderson's view of life: *"It is never too late to have a happy childhood."* At every age, Anderson had fun just living. Business is fun to him. Anderson's idea of having fun is meeting life head on—accepting adversity and change to his advantage. We have seen that as a child growing up during the depression, he had no toys, not even a bicycle. Whenever he has a chance, he seeks to remain young by having his own brand of fun finding new customers, handling credit problems with banks, and solving company problems.

Before we outline some of these guides in the form of quotations from Anderson that might guide others in achieving success as entrepreneurs, let us discuss the incidences, actions, and concrete evidence of Anderson's personal qualities. Many of the statements in words might sound like platitudes. Throughout the author's research, he continually observed actions and the perceptions of others about these actions that "spoke louder than words."

Old friends from the 1930s to 1950s often referred to the fact that Ralph Anderson was always interested in the other guy's problems and demonstrated that he was sincerely a friend. People have always confided in him because they trust him. It was this trust and sincere interest that made others want to be around him and to be associated with him in a business endeavor. This characteristic is so refreshing and unique in modern business that it might appear to be "homespun," "soft soap," and unbelievable. At first the author was wary of being taken in by an actor, or a smooth operator. Finally, he realized that this reaction was unnecessarily cynical and inconsistent with the facts as they are.

With every interview, Anderson was continually open with his answers and unhurried in discussions. If the author called on the phone, Anderson was always free to answer. Anderson provided all information requested, trusting the author so completely that the author felt weighty with responsibility.

One obvious omission in this book is net profits per year, return on investments, and pricing margins. Strictly speaking, in a private company, such facts are no one else's concern. Moreover, financial aspects are beyond the scope of this book; we can assume Belcan is a success.

Anderson enjoys reading self-improvement books, particularly those describing how to improve effectiveness in management. The principles he lives by are based on the fact that he likes himself as he is, and if the other guy doesn't, then that's just too bad. Another of Anderson's philosophies is that he strives to treat everyone as an equal. This is true for better or for worse; for example, he's not above admonishing an employee, and he'd quickly serve the same admonishment to his best friend.

In the early 1980s, when the company started rapid growth, Anderson wrote a set of principles. The increase in the number of employees made personal contact more difficult. He called this summary a Charter. The Belcan Charter was printed on a plastic stand and placed at each work station. See Figure 2.1.

Several observers point out that the Belcan Charter is quite similar to Deming's 14 Points. See Table 2–1. Interestingly, Anderson wrote his Charter in 1983, before Deming became well-known in the United States. At that time, in 1983, Anderson had no knowledge of Deming's ideas.

In an interest of instilling the ideas of personal development in each employee, Anderson publicizes a quote each day via computer mail. Now that time does not permit his talking with every employee every day, he still works to apply this theory of "management by wandering around," and communicating directly with as many employees as possible.

CHARTER

Belcan Corporation provides a full range of engineering services to clients in a wide variety of industries. The following set of principles will guide our business activities.

- Belcan is committed to excellence; dedicated to the customer and customer service and respect for the individual (Belcan's greatest assets are its people).

- Belcan employees should keep sales number one in mind, any action of any employee will affect sales.

- Belcan will be a market driven organization.

- To make Belcan grow, we will supply the tools to do the job and provide an atmosphere that encourages open dialogue.

- Belcan does good work and expects a fair price for our services.

- Belcan is a company that cares for its employees and clients.

- Listen to the customer, the customer comes first.

- Live by your word.

- With a positive attitude we will meet our goals, if we don't care who gets credit.

- If you make a mistake, admit it; if you don't know, say you don't know.

- Think positive about your work and life.

Ralph J. Anderson

Figure 2.1 Belcan Charter.

Table 2–1

14 POINTS FOR IMPROVING MANAGEMENT TECHNIQUES *By W. Edwards Deming*	
• Achieve constancy of purpose. • Learn a new philosophy. • Do not depend on mass inspections. • Reduce the number of vendors. • Recognize two sources of faults: Management and production systems Production workers • Improve on-the-job training.	• Improve supervision. • Drive out fear of failure. • Improve communications. • Eliminate fear of change. • Consider work standards carefully. • Teach statistical methods. • Encourage new skills. • Use statistical knowledge.

Basic Principles—Quotable Quotes for Everyday Living

Anderson enjoys talking with people about Belcan and about his personal philosophies. The following quotations have been collected from those made to the author during the many interviews, some frequently repeated. All of them are typical of "Andersonian quotes."

I can learn from anyone. This short quote describes Anderson's approach to people. His is the open mind to which he refers in the Charter. This simple idea not only makes it possible for him to gain new knowledge from his numerous personal contacts, but it renews life's challenges—often on a daily basis—and therefore keeps him young. As the company increases in size and resources, this message translates into continual contact with higher education, especially to personnel at the University of Kentucky.

Every setback in my life ended up being for my better good. This quote suggests more than just the positive thinking he refers to in his Charter. It refers to interpretations of past difficulties and seeing a silver lining in each; it is more than saying that everything will turn out for the best—it is saying that, in fact, all past bad occurrences really were for the good. He cites many examples of this conviction. Examples include his failure on air force tests as a pilot during World War II, and an automobile accident in Indiana in 1950.

Invest in the capital, land, and building before attempting to secure a contract for work. The client wants to see evidence that you are ready and able to fulfill your commitment. Anderson has always been ready to take the risk of obtaining facilities before a successful outcome was assured. He believes it is necessary to prove that you are ready and able to carry out promises. In fact, Anderson believes that the physical capabilities are more important than a signature on a contract. Actions and oral agreements speak louder than written words.

You must trust people. Anderson is always ready to deal with people on their word. He does not feel that an agreement in writing or a legal document adds much. If the person is not going to abide by his word, a legal support does not reinforce the moral commitment. It merely provides grounds for legal action. Such action takes time and causes one to lose track of the business at hand.

Give the other fellow the benefit of the doubt. People do not enjoy doing business with others who might cut corners. In the long run, the reputation of giving a square deal will pay off.

Management is people. This familiar quotation of Anderson did not come from studying books on human relations. To Anderson, the idea is pure common sense. Originally, Anderson was his only Director of Human Resources. Later, he added a competent Vice President of Human Resources, not only to reinforce his effort but also to protect the company legally with the increased laws dealing with human behavior such as affirmative action, sexual harassment, insurance, and the sunshine law.

Whereas some people promise to deliver services when needed, we deliver the services. Much of industry depends upon individuals meeting time schedules. Anderson's belief is that much success depends upon building the client's confidence that he would deliver—not deliver promises—but deliver the services. Good excuses are contradictions. Failure to meet a schedule is bad, regardless of the quality of the excuse for failure to deliver.

Go to work to have fun. If you enjoy yourself, you will do a better job. If you don't enjoy it, find another job that you can enjoy. This principle was first introduced in Chapter 1. Anderson continues to work to seek more wealth. He is so sold on business being fun, that he wants all his workers to have fun in their work. *A happy shop is a safer, more productive, better quality-producing shop.*

Innovations by Belcan outside of Engineering

Starting with Anderson's Charter, Belcan developed what in the mid-1980s became known as the Deming approach. Deming's approach had revolutionized Japanese management in the production of automobiles and other physical products. Since Belcan provided a service, it had to look for help in the service industry. Florida Power and Light (FPL) had been recognized widely as one of the firms that practices the Deming approach to quality as a service-oriented firm.

Belcan's service is even more nebulous than that of Florida Power and Light. Belcan provides *ideas*. At times these ideas are transmitted as drawings and models, yet, the engineering service remains elusive. The conclusion at Belcan is that its product is *customer satisfaction*. Certainly, a skeptic's first reaction is that this conclusion is another example of hype and public relations; it sounds good but is it a tangible, feasible idea? How else would one identify what is truly the company product: concepts, consulting reports, professional advice, and teamwork. Yes, after contemplation one can imagine that *Belcan Engineering's product is truly consumer satisfaction.*

Quality must be considered in everything the company does. Client satisfaction is the best measure of quality in a service company. Quality is a top requirement for each person in the organization, not just of the quality assurance specialist. The specialist can only train people to think quality one hundred percent of the time. He cannot put quality into the service.

The Deming system (see Table 2–1) provides the essence of quality focus. However, most of the system is directed at quality of products. At Belcan, *the product is service.* Service is much harder to measure as to its quality. Therefore, Belcan's Total Quality Leadership (TQL) had to be tailored to Belcan's situation. With Florida Power and Light as mentor, Belcan has contributed to the application of the Deming system in the service sector.

Total Quality Leadership has a number of specific details that have been explained in a company brochure. Quality is defined as "the measurement of results compared to mutually agreed upon expectations." The program encompasses strategic planning, production process, and continuous improvement.

Basic concepts of TQL are:

- *Training.* All employees are trained to think continuously of quality and satisfaction of the customers. This training is conducted not only by functional quality specialists, but each employee is empowered to teach and implement improvement.
- *Structured Steps.* The development of Total Quality Leadership uses a detailed program of steps with story boards, flow charts, and other check lists. These steps are kept simple and, through repetition, the idea of TQL becomes a habit by each Belcan employee. The most often repeated criticism of TQL is that it merely teaches the obvious, that it is routine and boring (since it strives to make quality a habit), and that it is simply a modern buzzword. Anderson, himself, responds

that he does not need to sit in training sessions because he states, "I *am* quality" and it teaches the common sense (gut feeling) basics that he has used since Belcan was a small firm.

- **Team Concept.** Employees contribute by means of the team approach. For this reason, top level officers are called the "lead team" and all employees are members of cross-functional teams that implement their engineering activities.
- **Communication and Understanding.** Even though quality involves technical aspects, each person should learn to communicate with clients and all persons affected by the process so that customer satisfaction is assured.

The foundation for TQL is in the implementation of the refined Continuous Process Improvement program at the project level. CIP is an ongoing education process that involves all Belcan employees. Its jargon has become commonplace: story boards, customer identification, quality evolution charts, collections of data on sub-activities plotted on Pareto Charts, and some of the Deming terminology.

In essence it is more than a work improvement or quality assurance program. It continuously focuses on identifying customers' expectations. It analyzes as part of the planning stage and provides well-thought-out goals and processes. It reduces costs.

The company set its goal to win the coveted Malcolm Baldrige National Quality Award. At the time of commitment, only a few companies had succeeded in obtaining the award. The company established a number of juried winners to "celebrate quality." Belcan uses formal presentations of monetary awards, photos, and certificates to continually remind employees of the central importance of quality. Each year a Quality Week Celebration emphasizes the steps taken by different departments to improve quality. The program, presented in the Anderson Way lobby, gives each department a means for showing in detail how they contributed to the overall quality program. The program is videotaped and is available for those who are interested in both Belcan's actions and its advanced approach to improving quality.

Results of applying CIP and TQL theory are impressive in improving the eight functions recognized by the Project Management Institute's Body of Knowledge (PMBOK): scope, quality, time, cost, risk, procurement, communication, and information. In projects of the industry involving the installation of a large 480 V substation above an existing warehouse, 26% of the project was spent on CIP before design and engineering started. Specific results included (1) the project started on time, (2) the design fee was not exceeded, (3) the mid-design changes were handled smoothly, and (4) only five field changes were identified. Field costs remained well within the Outstanding Goal.

In the 1980s Anderson recognized that he needed to think about improving his management style. He hired management consultants and read about the management philosophies of others. Whereas he had depended on his wife to provide management in the early days of the company, he now began to direct his innovative tendency to management. Anderson moved so rapidly until 1989 that he never really considered management principles and practices.

After a serious financial crisis, he reorganized (that is, he released a number of executives) and allowed key executives who had thoughts of formal organization to apply their knowledge as presidents of two subsidiaries: J. J. Suarez and Lane Donnelly. It is interesting to note that the new president of engineering services is an engineering Ph.D. and the new Vice President of Human Resources is a professional psychologist.

Until this time, planning in the formal sense was ignored. Control systems in the form of engineering instrumentation were prepared for clients; they were not used as internal management control systems within Belcan.

To Anderson, leadership is a simple idea: hire good people whom you can trust, work with them as hard as anyone in the organization, encourage openness and free flow of direct personal information, motivate fellow workers by showing them how much fun working at Belcan can be, and when a new idea comes to mind by anyone, let them "give it a go." He believes that compensation is directly related to quality of effort and results. Between 1989 and 1991 the company's direction was returned to Anderson's tried and true approach integrated with Suarez's strategic planning and straightforward charting.

Religion

Anderson grew up in a Baptist family and Ruth Tucker grew up as a Brethren. After they were married, they chose the Methodist Church. Their daughter, Candace, was a factor in the Anderson family joining the Methodist church.

Religion to Ralph Anderson is a personal matter; he is not a weekly church-goer. In fact, he identifies with those preachers whose theme is closest to his personal belief. The writings and sermons of Norman Vincent Peale fit his personal beliefs. He buys books by Peale and other authors who stress positive thinking. Several copies of his favorites are available so that he can give or lend them to Belcan employees and friends.*

After they bought Mercer County farms, the Andersons drove weekly from Blue Ash to Mercer County with at least four hours driving time per week. Anderson developed a routine of listening to audiocassettes as a means of focusing on religious and mental development during the weekend drives. Built into this routine was the normal plan to arise early on Sunday morning in Mercer County, drive back to their home in Indian Hill, listen to audiotape en route, and arrive at home in time for the weekly telecast of Reverend Robert Schuller from the Crystal Cathedral in California. Peale's "positive thinking" and Schuller's "possibility thinking" fit nicely with some of the Anderson quotations earlier in this chapter.

Throughout this book we have described Anderson as perpetually on the outlook for new technology, new approaches to management, and even to new interpretations of history. In the area of religion and expanding the possibilities of the mind, Anderson reaches for ideas of life. For example, in 1990, Ralph and Ruth were initiated into Transcendental Meditation (TM) and tried the twice daily sessions of meditation according to Maharishi Mahesh Yogi. Like he does with favorite books, he purchases several copies of audiocassettes and CDs of his favorites and listens to the recordings during the weekly drives to the Kentucky farm. Some of these recordings include *Magical Mind, Magical Body*, by Deepak Chopra, and *Head First: The Biology of Hope*, by Norman Cousins. Catalogs of many tapes by Nightingale Conant, Anthony Robins, Wayne Dyer, and others supply Anderson with a wealth of ideas to assimilate during his weekend drives. The books, recordings, and videos of Dr. Robert Schuller supply Anderson with a wealth of ideas with which to help others. (See Appendix C for a list of these books.)

During the five-day week in Cincinnati, at an age at which others are retiring, Anderson keeps a schedule that begins at 5:00 A.M., including some working breakfasts and snack lunches at his favorite lunch spot. He does not play golf or have any other recreation other than his relaxation over the weekends on his farm. Even during the "relaxation weekends" he spends several hours discussing business with Harvey Mitchell, his farm manager.

Anderson repeatedly states that without the weekends on the farm his life would be cut short. His activities and comments about the farm visits

*Wayne W. Dyer, *Your Erroneous Zones*, Funk & Wagnell, 1976; Claude M. Bristol, *Magic of Believing*, Pocketbooks, 1949.

provide illustrations of his relaxation. His observation of nature, e.g., watching birds and groundhogs, and his appreciation of natural beauty indicate that he can accelerate his restful responses. Since he enjoys all types of activities he tends to regenerate his capacity for strenuous work with minimum stress.

The interesting underlying concept repeated throughout the life of Ralph Anderson and through the history of Belcan is the simple idea that being an honest guy who has fun at work without formality and acting will pay off in increased sales and improved quality. His secret is really no secret. His formula is a collection of unsophisticated and simple ideas.

Anderson's achievements in applying this philosophy received recognition on May 8, 1994, when he received an honorary degree, Doctor of Engineering, from the University of Kentucky. The citation supporting this award emphasized Anderson's lifelong service to the Commonwealth of Kentucky and its state university. We shall see in later chapters that Anderson has received many other awards; however, the honorary degree from his alma mater was the crowning honor to a happy life of service to his home state.

EDUCATION AND PERSONAL DEVELOPMENT, 1923–1957

Anderson's Life-Long Kentucky Identification

"As a boy, I sold sheep and so they called me buck or bucksheep."
RALPH G. ANDERSON

T The life of Ralph Anderson intertwines with the history of Kentucky. Although he emigrated to Ohio after his education, like many of his fellow Kentuckians, his roots in Harrodsburg and Mercer County remained a strong influence throughout his entire life.

He continually demonstrates his loyalty to Kentucky and its state university. As the facts of his early life, his education in Kentucky, and his continual return to his native state are outlined, we observe how his heredity as a Kentuckian and the environment of central Kentucky provides a base for analyzing the development of this entrepreneur.

Anderson's birthplace is Harrodsburg, one of the oldest settlements west of the Alleghenies. Those pioneers who travelled over the Cumberland trail to settle in Harrodsburg were people who preferred the independence of frontier life and who ventured into the wilderness in place of the stability of cities and towns of the east coast. The opportunities of the frontier environment encouraged entrepreneurial effort and independent thinking about life. One well-known group of settlers was the Shakers who founded Shakertown in Mercer County circa 1820 within a few miles of where Anderson grew up and later built his large farm. The innovations introduced by the Shakers were forerunners of the high tech thinking of the 1990s. Four former governors of Kentucky had their roots in Mercer County: Christopher Greenup, 1804–1808; Gabriel Slaughter, 1816–1820; John Adair, 1820–1824; Beriah Magoffin, 1859–1862. Heredity and environment supported entrepreneurs.

This publication follows the bicentennial celebration of Kentucky becoming the fifteenth state of the United States. An Encyclopedia of Kentucky published in 1992 offers descriptions of the people of Kentucky with whom Ralph Anderson identifies.

Anderson relates major events in his life to his attendance at athletic events of the University of Kentucky. In the early recruiting for Belcan, graduates of the University of Kentucky were considered with great favor. Later, when he had time to study the educational history, he took pride in emphasizing the importance of John Bryant Bowman, an alumnus of Bacon College located in Mercer County, to the creation of the University of Kentucky. Bowman was regent of Kentucky University (1865–1878) and provided the vision for a great university in Central Kentucky. As a loyal alumnus of the College of Engineering, Anderson submits that Henry H. White, who had studied civil engineering at Bacon College, was actually the first alumnus of the College of Engineering. Although historians do not share that view, Anderson enjoys the thought that Mercer County was very closely related to the present University of Kentucky.

Anderson publicizes his Kentucky roots by flying the state flag of Kentucky alongside that of the Ohio and the American flag. (See Colorplate 4.) His investment in farmland in Mercer County from 1967 through the 1990s further cemented Anderson to his roots. His purchase of several historic large Kentucky homes built in the early part of the nineteenth century caused him to study the early history and architecture of his native state. For over a decade, the Andersons returned each weekend to their property in Mercer County. (See Appendix A for a detailed description of activities on the Anderson Circle Farm in Mercer County.) Anderson was not alone among Kentuckians who ventured north into Ohio hoping for a better living only to be drawn back to their native state, which they loved and could not leave permanently.

Family Ancestry

Ralph Gilbert Anderson was born on a farm two miles from Harrodsburg on July 19, 1923. His father, William Robert Anderson (born July 19, 1888) and his mother, Mattie Jane Cunningham Anderson (born August 6, 1892) were independent farmers of modest means. Ralph was the youngest of four children, with a brother, Frank, and a sister, Roberta, who were so much older than Ralph that one time after Ralph was grown, he spoke to Frank only to find that he had to tell his brother who he was. Gladys Dean is an older sister by three years. She remains in Harrodsburg, working at the Mercer County Courthouse.

The lineage of Anderson through his great-grandparents (Cunninghams, Wrights, Cozines, Downey, and Cooverts) includes families who have been in Kentucky since circa 1800. Most came directly into Mercer County from the east coast. In one case, after Thomas Anderson married Sarah McCarty in 1801 in Shelby County, the couple immediately moved to Mercer County. By the mid-1800s, all ancestors of Ralph Anderson's were natives of Kentucky. The families came to Kentucky from New Jersey, New York, Virginia, and North Carolina. Many were from the Netherlands and others from Scandinavia.

Some interesting observations by Jon P. Neill, Cincinnati, the genealogist who traced the Anderson family, showed the Van Cleve family connected to Anderson, the Wright Brothers, and Daniel Boone.

> The Van Cleve family is truly a backbone of American pioneerism. The seventeenth century in New Netherlands was a wilderness with untamed land and people, as was nineteenth-century Kentucky. The true American pioneers did not begin in the western states, but rather in the unsettled lands of the East. The family lent many of its men to the cause of the American Revolution. The Boone family was closely allied to the Van Cleve family, and today their descendants claim a common heritage. In peacetime, the Van Cleves provides clergy and innovators. Orville and Wilbur Wright share with many other families the distinction of being descended from a remarkable group as the Van Cleves.
>
> Jane Van Cleve married Squire Boone (brother of Daniel Boone). Ralph Anderson is second cousin, five times removed from Jane Boone . . . Anderson is a sixth cousin of Orville and Wilbur Wright through the line . . . back to Isabrant Van Cleve (Wright's first bicycle was called the 'Van Cleve').*

Profile of an Entrepreneur

What are the qualities which distinguish an entrepreneur? Entrepreneurship is more than a professional orientation; it is a matter of attitudes and approach to life. Whether one takes the position that heredity and genes are important or that the environment conditions the individual, both the

*Jon Patraic Neill, *The Ancestry of Ralph G. Anderson,* December, 1988, unpublished (but typed and bound), Abstract Seven, Report Three, pp. 9–10.

lineage of Anderson and the culture in which he grew up provide reinforcement to the idea that entrepreneurs develop from programmed genes and the climate in which they are nurtured.

Early Childhood

Anderson's early childhood was spent in a small frame house on Warwick Road with minimum conveniences and coal heating stoves. The house was at the edge of town and offered space for a small farm with the usual chickens and a cow. Ralph was expected to help with the chores; his father attempted to teach him to milk the cow. Ralph did so poorly that he remembers vividly his father saying, "You can't even milk a cow—you will never amount to anything." Because of this comment, Ralph visited a neighbor across Warwick Road. John Deshazer had a farm of more than 100 acres. Ralph secured a job at $1.00 per day, a significant sum of money for a young boy at that time.

The job with Mr. Deshazer gave Ralph cash, independence, and a chance to demonstrate to Deshazer and other neighbors along Warwick Road that he was a hardworking, honest person. The seeming incidental comment by his father about ability to milk a cow had strategic significance for the formation of Belcan Corporation.

For a person who desires to form an independent firm, the usual roadblock is financial backing. Ralph had no contacts with banks or wealthy friends. Ralph did impress his friends and rural contacts that he was honest and a good risk. His open personality served to help at critical times in obtaining needed loans. Another neighbor on Warwick Road, Ott Elliot, later made a crucial loan to Ralph when other funds were unavailable. In fact, at critical times, even after banking lines of credit had been established, Ralph found his friends not only willing but eager to help him meet a crisis.

Ralph's father had several jobs to supplement his farm income. In the winter, he was manager of the Farmers Tobacco Warehouse, and later became an employee of the Kentucky Utilities Company. His father was affectionately known as "Mr. Bob" and had earned the reputation of an honest, hardworking citizen of Harrodsburg. The Anderson family remained too busy to develop contacts with the well-established old families of Mercer County, but whenever given a chance, the Andersons were respected in the community for their trustworthiness.

Ralph attended the Harrodsburg elementary school, the "school on the hill." His classes had 30 students, which tended to overcrowd the classroom.

He had several close friends but much of his spare time was spent in helping at home. One classmate recalls that Ralph was a joker and enjoyed teasing girls. By and large, he was a boy's boy, interested in athletics. Until he left Harrodsburg for a job in Ohio, he was so busy working, going to school, and playing football that he could not get to know many of the old families of Mercer County.

In the early 1930s during the Great Depression, Ralph first demonstrated the initiative of an entrepreneur. Neighboring farmers grazed sheep. At times, a ewe would not accept a lamb for nursing. In such cases, the lambs had to be fed with a bottle. Ralph was able to obtain these lambs, nurse them with a bottle, and sell them. This example explains why his friends gave him the nickname of "Buck" or "Bucksheep," a name that stuck with him during secondary school.

High School

Interviews with classmates consistently elicited responses that Ralph had an "outgoing laugh," but ironically he was neither an extrovert nor an introvert. People liked to be around Ralph throughout his life. Evidence of the solidity of friendship from high school days appeared in 1988 during a class reunion of the 1938 high school class.

Ralph earned the reputation by his high school days as being a hard worker and always doing his best at whatever he tried. In school he was an average student and didn't appear to be exceptional except in mathematics classes. He was smaller than others of his age, but as a member of the Harrodsburg High School football team, he filled a guard position while his friend John Harris, who moved to town from a farm near Shakertown for the 7th grade in 1937, played in the backfield.

First Job Away from Kentucky and into the Armed Forces

In September of 1941, after graduation from Harrodsburg High School, Anderson moved to Cincinnati to work with the Curtiss Wright Corporation. His close friend, John Harris, had already gotten a job at the aircraft company, cutting fins on air-cooled engines. Anderson's job was cutting gears. After using the method that had been previously taught to new employees, Ralph developed improvements in the gear-cutting process that improved

Harrodsburg Class of 1938—Photo includes several of Anderson's friends who supported him at critical stages in the development of Belcan. Both Ralph and friends have remained interested in one another throughout the years.
1st Row, Left to Right: *Billy Smith, Lyda Cornelius, Roy Sallee, Archie Martin, Leslie Rue, Ralph Anderson, Billy Bob Edger.* **2nd Row, Left to Right:** *Frances Draffen, Andora Brown, Chester Scott, J. W. Keebortz, Ernestine Reed, Beuna Milburn, Pauline Webb, Johnny Harris, Haldon Yates, Lois Gordon, Dorothy Lay, Clarence Harris.* **3rd Row, Left to Right:** *Mary Katherine Hopper, Charles Noel, Frances Board, C. B. Yates, Virginia Balden, Roy Scott, Marcia Sexton, Jimmy Keightley, Valoise Terhune, Henry Davenport, John Sullivan, Jason Bugg, Johnny James, Billy Goddard.*

the quality of the gears. This first job in the aircraft industry introduced Anderson to the new developments in that industry. This proved later to be of major importance to him and the future Belcan; the fact that General Electric was located in Evendale, Ohio, and Allison Gas Turbines was located in Indianapolis, were key determinants of Anderson's later success.

Military Experience

Though Anderson had registered for the draft, in March of 1943 he volunteered for the Air Corps. During that time, the Air Corps was part of the U.S. Army.

His first training was in Illinois but later he went to San Antonio, Texas for flight training. During the final days in flight school, he was informed by the instructor that he did not qualify; Anderson failed in his effort to become a pilot.

Anderson was disappointed by his failure to achieve his goal. Yet, it was during this time in his life that he developed a fundamental philosophy that things happen for the best. Even though results appear to be negative, Anderson repeats throughout his life: "Everything happens for the best." His philosophy is continually reinforced. For example, the failure to make flying status was a shock to the young Ralph Anderson. However, after the war he discovered that every other cadet named Anderson who had been in flight school with him was later killed during the war. Ralph was the only Anderson who made it to the end of the war. Ralph's last job in the air corps was as a flight engineer on a B-29.

University of Cincinnati

With no money to attend college, Anderson took advantage of the GI bill to help pay for his education. In January, 1946, he entered the University of Cincinnati under its cooperative program so that he could work for an industrial firm while taking formal courses. The program offered a chance to earn money and provided an opportunity for practical experience. However, it required the student to attend college for five years instead of four.

For the first two years Anderson took the basic engineering courses at the University of Cincinnati. In addition, he worked in material handling and in the maintenance department of American Rolling Mills (later, Armco) plant in Middletown, Ohio.

In January of 1948, Anderson decided to return to his native state to complete the degree in mechanical engineering at the University of Kentucky. The in-state tuition afforded a Kentucky resident offered important financial savings. He moved to Lexington and shared a room with friends on East Maxwell Street.

Courtship and Marriage to Ruth May Tucker

At Armco, Anderson met Ruth May Tucker. Ruth Tucker lived with her parents on Sherman Street in Middletown. She had a bookkeeping job in the canteen of the American Rolling Mills. After several months of dating, Ralph and Ruth decided to get married.

On several occasions, Ralph and Ruth had visited John Harris. Harris had married his boyhood sweetheart, June Solver, in Lexington. Ralph and Ruth asked John and June to help arrange a wedding in Lexington. At the time, John and June had an apartment at 242 S. Limestone that was convenient for John to work at Wilson Machinery and June to work at Western Union in downtown Lexington.

John Harris checked with the minister of the First Methodist Church around the corner from his apartment and arranged for the wedding. The minister of the church performed the ceremony on March 27, 1948 with the Harrises as the only witnesses. Furthermore, since Ralph did not have an automobile, the Harrises lent their car to the newly married couple for a week for their honeymoon. During this week, the Harrises walked to work.

The Andersons returned from their honeymoon and rented the same apartment that the Harrises had lived in. John Harris had decided to move to Nicholasville to work in an auto repair garage. The Harrises, however, kept in touch with the Andersons throughout their lives. Both couples felt that friendships were important, with Ralph and John remaining friends from the seventh grade in Harrodsburg.

Mechanical Engineering Degree from the University of Kentucky

In Lexington, Ralph immediately obtained a job in the machine shops of the College of Engineering of the University of Kentucky. The University of Kentucky did not have a cooperative program but they were able to provide income to some students who wished to work long hours assisting with the college activities. Ralph also still had the GI bill to help with finances.

Soon after their marriage, Ruth was able to find a bookkeeping position at the Southeastern Greyhound Lines, headquartered in Lexington several blocks from the Limestone apartment. During these first two years of marriage, both Ralph and Ruth worked long hours. Ralph often worked at nights in the machine shops. The classes and labs usually took most of the days.

As Ruth kept books for the bus company, she naturally acquired the role of financial specialist for the couple. She kept the checkbook and managed the cash flow. From the start of the marriage, the couple decided that they would maintain a joint bank account even though each had a separate income. Ruth was more interested in thinking of every detail of expenses and Ralph preferred to keep approximations in his head and not worry about the details of bookkeeping. This preference for keeping approximations in his head remains strong today.

Undergraduate Studies

In his undergraduate work, Anderson was viewed by the engineering faculty as an average student. Yet he worked hard and he was popular with students and faculty. He took the maximum number of college credits during each semester even though he continued to work in the machine shop. Ralph progressed rapidly toward a degree. However, since he transferred credits from the University of Cincinnati to the University of Kentucky, he was required to take several additional courses to complete requirements at the University of Kentucky. Because of these extra required courses, Anderson did not graduate until August, 1950.

During the final term in the summer of 1950, Ralph earned fifteen credit hours. The usual maximum was twelve hours. To qualify for summer graduation, it was necessary for him to use correspondence courses in which a student studied independently, completing daily assignments and taking a final examination. Because Ralph was so busy, Ruth typed the daily assignments on an old L. C. Smith typewriter. Ralph admits that Ruth secured A's on the daily assignments whereas he only obtained a B on the final examination.

Ruth's old L. C. Smith was a rented machine. When she attempted to return it, the owner of the store informed Ruth that the machine had little value and told Ruth to keep it. The same typewriter was used for three decades afterward during the growth of Belcan and remains in the Andersons' Indian Hill home yet today.

Jobs with Three Large Corporations

Barber-Coleman Corporation

Upon nearing graduation, Ralph sent out 100 resumes. From these, his only offer was from the Barber-Coleman Corporation in Rockford, Illinois. It was September of 1950 and he accepted the position immediately. The position was in the aircraft controls department, working on subcontracts for the aircraft industry. Ralph's pay was $180 per month. Once again, Ruth obtained a bookkeeping job.

Ralph and Ruth rented a small apartment. The kitchen was closet-sized and they shared a bathroom with other tenants. Living conditions were minimal.

Though Anderson was situated some 400 miles away from Lexington, he still enjoyed attending football games at the University of Kentucky. In the fall of 1950, he was headed through Indianapolis when he suffered an automobile accident.

General Motors

Considering the conditions of his job in Rockford, Anderson pursued a job in Dayton, Ohio with the Frigidaire Division of General Motors. The auto accident added to the development of Anderson's optimistic philosophy: "Everything turns out for the best in the long run."

Of particular note is the fact that Anderson made the decision to work for Frigidaire very quickly. His typically analytical mind does not prevent him from making major decisions within minutes. It may be referred to as decision by intuition; Anderson calls it decision by gut feeling. The self-confidence exhibited in decision-making abilities singles out analytical thinkers. This is a distinct advantage when negotiating. Of course, if there were more incorrect decisions than correct ones, and disastrous results, the analytical individual may lose self-confidence and therefore, the ability to make snap decisions.

So the Andersons relocated, this time to Dayton. And once again, Ruth located a bookkeeping job with a paper company.

Ralph's job was associated with the development of positive-replacement reciprocating compressors, controls and switches for air conditioning and refrigeration units. Since the daily work required little extra effort, he considered it routine, boring and lacking in motivational challenges. Nonetheless, Ralph stayed with General Motors for two and a half years learning more about testing engineering equipment. This illustrated a second important Anderson philosophy, "Something can be learned from anyone and from any situation." Although the job was boring, Anderson recognized the potential learning opportunities and he stayed with it. Moreover, there was the added advantage of making contacts while at Frigidaire and other large firms. These contacts became a major asset when he decided to form his own company.

General Electric Aircraft Engines

In the mid-1950s, Anderson moved from Dayton back to Cincinnati to accept a position with General Electric at Evendale working on gas turbine aircraft engines. Ruth secured a job in Cincinnati.

Ralph enjoyed working with GE for the first year. Shortly after that, he became dissatisfied, feeling the company was not treating him properly. One morning he took it upon himself to walk into the office of the vice president and general manager, Cramer W. "Jim" LaPierre to express his concerns about his job. Another element of Anderson's philosophy, this one specifically work-related, was developing. He considers it easier to talk with the highest official available than with subordinates.

Entrepreneurial Musings

It was during this period with GM and then GE that Anderson began thinking about the possibility of going into business for himself. He had noticed that his brother-in-law, Dan Tucker, who had an auto shop, could compete and sell even without the precision of fine tolerances.

A basic operating principle began to formulate in Anderson's mind: "people management." That is, real management involves dealing on a person-to-person basis and not on the bureaucratic level of one department to another. The heart of management is in the people, not in organizational departments.

In 1953 Anderson realized that he would be happier working with a smaller firm. He felt that innovative ideas had a better chance of being listened to in such a small firm. Furthermore, an entrepreneur in a small firm would provide him with a greater learning experience about running his own small business.

Kett Corporation: Role Model for an Entrepreneur

Anderson contacted Kett Corporation to inquire about work. Kett was an engineering specialty firm. During his telephone conversation, the person on the other line responded, "Ralph—don't you know who you're talking with? This is Doc Savage!" Savage had been one of Anderson's professors in the College of Engineering at the University of Kentucky.

Kett had about 70 engineers and six project managers. Anderson was hired in 1953 as one of those project managers. His duties involved work in stress analysis and heat transfer, working on such programs as the J-47 engine and the B-47 installation design.

Though Ralph was located physically at GE, his organizational position was project manager with Kett. His employer, Karl Schakel (See *Profile: Karl Schakel*) had a management style that made Anderson more satisfied because he had greater freedom to use his skills and creativity.

Anderson developed a reputation in the area of stress analysis in turbo jet engine frames. He had the opportunity to apply these skills with other firms. One of the later assignments with Kett at GE was on heat transfer for GE's aircraft nuclear propulsion department. It was here in the mid-1950s that Anderson met an unusual manager, Jack Hope. Anderson and Hope were to remain friends over the next forty years. (See *Profile: Jack Hope*.)

These three years spent with Kett were probably the most significant in preparing Anderson for the formation of Belcan. There are three main reasons for this significance:

- Anderson learned how a small technical service engineering company operated.
- Karl Schakel was an excellent role model for a budding entrepreneur. Schakel was not aware of the significance of his relationship with Anderson at the time. Later, Anderson recalled the details of how important Schakel was to him during this key period.
- Anderson worked for Jack Hope on a most challenging project—aircraft nuclear power. Anderson and Hope were quite different individuals but each was impressed by the other. It takes a keen mind to recognize another and perhaps it was the differences in style which each cherished about the other. The paths of Hope and Anderson were to parallel and periodically join years later.

Schakel recalled in a 1991 telephone interview with the author that Anderson was a hardworking, competent engineer, but he did not truly appreciate him until Anderson informed him that he would not remain with him when he moved to Florida. Anderson prefers to stay close to his home state.

In May of 1957 Schakel sold Kett to U.S. Industries and moved to Florida.

Before Schakel left, he and Anderson discussed the fact that Ralph had developed excellent contacts with Allison Gas Turbines and General Electric. Ralph asked if it would be all right if he were to follow up and make use of these contacts on his own. Schakel had no objection, believing these contacts would present no conflicts with his current interests. During the next twenty years it was these contacts which proved to be intangible assets for the new company. The major drawback was that banks would not lend money using the contacts as collateral.

Even though Schakel moved physically out of the life of Ralph Anderson the impact of his thinking had found fertile cells in Anderson's mind. All of the tendencies toward entrepreneurship were present in Anderson from boyhood. All that was needed, to use aircraft engineering jargon, was a short thrust from an afterburner (Schakel as role model) to push Anderson away from working for other corporations to forming his own firm.

Starting a New Business—Conditions Bad, but Anderson's Intuition Is to Move!

Anderson had purchased his first home at 10285 Pendery Drive in 1956, moving from an apartment on Schenk Road in Deer Park. Candace, an only child, was born on March 1, 1956. Anderson had no capital and, in fact, had gone to his old friend in Lexington, John Harris, to borrow approximately $1,000 for living expenses because he had no credit rating at banks. Harris and his wife June cleaned out their savings, including their piggy bank, and gave John's buddy cash, not asking for interest. In spite of the fact that Anderson had no funds to invest, the idea of forming his own business continued to develop. He worked for Kett and later, Ketco to provide income.

Ketco, 1957–1959

Schakel had found that pricing and handling accounting differed between dealing with the government and private firms. He formed Ketco Corporation to specialize in contracts for the Federal Government while Kett continued contracts with private firms.

Anderson remained at Ketco in Cincinnati. While at Kett, Anderson had met Bob Mendenhall who became President of Ketco. Anderson became sales manager and served in that role until 1959. He quit at this time because he felt constricted; in comparison, he had been contented at Kett because the company had provided room for his innovative tendencies. Anderson preferred flexibility and freedom.

In the early 1960s Schakel sold Ketco to Waltham Watch, which later changed the company to a temporary services agency.

Bob Mendenhall was another person who was pregnant with ideas but, for one reason or another, they rarely worked out. One of these ideas was known as Aero-Math. While at Kett, Anderson and Mendenhall began work with Aero-Math, but after two weeks, Anderson called it quits. Anderson was always willing to listen to ideas of others and often willing to take a risk to test out the prospects. However, another skill which he developed early (and would call upon later) was the ability to recognize the ones that would lose money and to get out before he could lose even more.

In working with Karl Schakel and Kett, Anderson learned the basics for operating an engineering specialty firm.

Stage Set for the Formation of Belcan

It has been said that no time is optimum for getting married, having a baby, or forming a new company. The best of times and worst of times for Anderson to dive into a new enterprise was the year 1958.

The economy was in good shape. After Sputnik, engineering was the "in" profession. American businesses were showing the world "the American way" in competition with communism. Cincinnati remained a center of industrial technology. Anderson had developed a good name with potential clients.

But Anderson had no money for capital investment. He had a family who needed a steady source of income. He had no "track record" except as a professional engineer. He had many friends but none had money to be his sponsor. *For entrepreneur Ralph Anderson, his gut feeling told him that this was the time to move.*

After leaving Ketco, Anderson had no income. One way to state the situation for Ralph, Ruth, and Candace between 1959 and 1961 would be that things were bound to get better, because they certainly could not get much worse. Ralph always retained his optimism—an essential ingredient for an entrepreneur!

PROFILE OF KARL SCHAKEL

 Karl Schakel was the role model for Ralph Anderson in the creation of Belcan Corporation in 1958. Books about how to become an entrepreneur may not be effective, but a living role model can demonstrate the actions of an entrepreneur, and this book on another entrepreneur can substitute for a living role model.

Karl Schakel was born in Cincinnati in 1921, receiving his engineering degree from Purdue University. He moved to California to work on new technology of the afterburner in the J-71 engine for the Phantom airplane.

With his experience in California, he returned to Cincinnati and formed the Kett Corporation in 1951. The name Kett was from Schakel's middle name, Koett. Kett Corporation was an engineering services firm having projects with General Electric, Cincinnati Milicron, and other firms.

When comparing entrepreneurs, Schakel's story is an interesting one. He began as an entrepreneur at the age of 30 (Anderson was 35 when he founded Belcan). He became known as the "cowboy capitalist." In 1970, he formed Western Land and Investment Corporation, Western Agri-Management Internationals, Inc. Out of Fort Collins, Colorado, his company helped change the appearance from the air of several states in the West with the circles of green formed by irrigation sprinklers. Operations extended to four continents and included helping third world countries. In 1990 Anderson noted in a list of 50 persons who knew him best, "Schakel got me in the engineering business; taught me the business in the 1950s."

Like Anderson, Schakel gives no indication that he will ever stop creating new ventures. The chief difference is that Schakel works primarily in the western half of the United States whereas Anderson concentrates in the eastern half.

PROFILE OF JACK HOPE

Jack Hope has remained a most significant friend of Ralph Anderson for 40 years. The two men were interested in aircraft engines; both liked to try new technology. Anderson thrived in taking financial risks; Hope thrived in experimenting with new things for large companies and for the government. Hope would be known as an "intrapreneur"; Anderson as a entrepreneur.

Hope might have ventured into new companies but for the opinions of his wife Anita, whom he loved deeply. Anita was conservative and discouraged Jack from taking risks. Ruth encouraged Ralph to do what he wanted to do and supported him in everything even though she would give him independent opinions.

The relationship of the two men is difficult to generalize. Hope has a creative mind which challenges Anderson by his rapid-fire thinking on technical matters. Hope provided ideas that helped satisfy Anderson's quest for having fun in creative thinking and developing new things. Hope and Anderson maintained an informal partnering relationship. Each worked for the other and each worked jointly in a small corporation, Haeco, Inc., chosen from the initials Hope, Anderson Engine Company. Hope never did have a managerial position formally in Belcan although he was a top executive in three large corporations and served as a science advisor to the White House and in other capacities of the federal government.

Hope had entrepreneurial characteristics but tended to move with impatience from one major position to another in several large corporations. He

JACK I. HOPE
HILLSBORO, OHIO

seemed to want to be so independent that he did not want to follow through with decades of building a single company. He had national contacts and was able to continue to use these contacts from his base on his farm in Ohio.

Anderson and Hope retain strong personal respect for each other but both are quite happy to go his own way. Periodically, Hope will appear with new ideas and energy.

Hope's personality and intelligence leaves anyone who talks with him exhausted from innovative thinking. The contacts are most stimulating but frustrating. Hope continues to be an "intrapreneur" (innovator within a large firm) instead of taking risks associated with striking out on his own as Anderson did.

Hope was born on a farm near Ramsboro, Ohio, on June 11, 1928. He attended Ohio public schools, and after World War II, he graduated from the Catholic University of America, in Washington, D.C. with a Bachelor in Aeronautical Engineering in 1951. During college he worked at odd jobs. In one of these he chauffeured the president of CBS to Truman's inauguration, and later he drove Mr. Rulling, Vice President of General Electric, a part-time job that provided the opportunity for a professional position since Mr. Rulling encouraged him to apply for an opening at General Electric.

Hope began twelve years with GE at Lynn, Massachusetts working on aircraft engines. He immediately met the legendary Gerhart Neumann. His relationship opened up interesting assignments for Hope including experiments in the nuclear aircraft engine program at GE. Neumann served as Hope's mentor.

In 1952 Hope was transferred with Neumann to Evendale works in Ohio. Soon Neumann was sent back to Lynn for a major development and Hope, as a young man, was left as Manager of the Nuclear Propulsion Department in Evendale. It was in this position that Hope first used Anderson in testing aircraft engines. Anderson tested the J-47 (and the related X-39). After that time Hope and Anderson remained close friends.

In 1960 Hope joined North American Aviation in California as chief of Propulsion and Thermodynamics where he managed personnel in support of such programs as the B-70, X-15, Apollo Support, and the B-1.

After four years, Hope returned to the Midwest as Vice President of Performance at Cummins Engines in Columbus, Indiana with 1400 people reporting to him. Experimenting on his own time on an engine, Hope called Anderson asking for two engineers to help him develop a revolutionary engine. Robert D. Johnston, one of the two engineers sent by Anderson, conceived of refinements in the engine. Engine Systems, Inc. was formed, and in 1968 Hope commuted from his farm at Hillsboro, Ohio to Belcan to work on the engine later known as the giesel. Development contracts with the U.S. Army provided the funding. Anderson obtained the patent rights on the giesel in 1970. A model of this revolutionary engine is exhibited in the entrance lobby at Belcan on Anderson Way.

In December of 1971, Hope moved to Washington to become Assistant Science Advisor in the office of Science and Technology in the White House for President Nixon. In March of 1974, Hope moved from the White House to GE to manage the CFM 56 joint venture with SNECMA, a French company (the CFM later became the most successful aircraft engine in the world, powering Air Force's KC 135, Boeing's 737). From October 1977 to 1982 he served for GE on a part time basis in exploring the possibility of establishing an international joint venture for developing an SST.

During the 1980s Hope was corporate director and consultant to Nadite, Inc. From time to time Hope performed engineering and management work with the U.S. Government, Battelle Laboratories, Caterpillar, Inc., Teledyne, Continental Motors, and many other large companies. He presented papers for the U.S. Air Force Scientific Advisory Board, ASME, AIAA, and NASA. He also returned to work with Neumann and GE on projects for the U.S Government and even after the patent expired on the giesel, Hope continues to secure development contacts.

Hope's wife Anita died on February 8, 1984, and Jack returned to his Hillsboro farm where he spends most of his time.

BELCAN: TEMPORARY ENGINEERING SERVICES, 1958–1975

"I discovered that I was quality, so I formed my own company."
RALPH G. ANDERSON

T he incorporation of Belcan in 1958 resulted from Anderson's aspiration to have a business of his own. The name Belcan was derived from the first three letters of the name of Anderson's daughter, Candace, placed after the first three letters of her friend, Belinda Beckett, who was the daughter of an Anderson colleague at Kett.

Belcan's initial operations were first located in a one room office at 9505 Montgomery Road. Anderson's job at Ketco provided a living for the family until Belcan's own business could be developed.

At the time of establishing Belcan, Anderson had no line of credit at the bank. Since he had little savings and no credit standing, he borrowed a small amount on his life insurance and again turned to friends in Harrodsburg, Kentucky. Ott Elliot, a farmer, had been a neighbor on Warwick Road as Ralph was growing up. Elliot had hired Ralph to work on his farm; Anderson's reputation as an honest, hard worker encouraged Elliot to lend $7,000 to him. Elliot said that he would not charge interest. However, since Anderson knew that Elliot was a religious person, loyal to his church, Anderson told Elliot that he would make contributions to Elliot's church equal to the interest that would have been due. This pleased his friend.

Anderson had four important advantages at the start of Belcan:

- A good reputation with engineers from his previous work.
- Continual optimism carried him through rough times. He took adversity as a challenge. In his mind, he made each misfortune a time of opportunity.

41

- He felt that he did not fit into larger corporations where chances for advancement appeared too slowly; by "doing his own thing," he had more control over his future.
- Every time he worked on an engineering job, he developed ideas for improvements. He had noticed that after World War II several engineering firms concentrated on providing temporary services to industrial firms. This concept was the original mission for Belcan and remained the most stable throughout.

Since Anderson had few assets, meager savings, and minimal support from city commercial banks, the above advantages proved to be critical at Belcan's inception and remained very important throughout its existence. From the outset, Anderson and Belcan necessarily depended on human services as the central theme. Without savings or capital, this theme predominated: MANAGEMENT IS PEOPLE. Anderson's direct, personal approach to business problems appeared a natural advantage over the bureaucratic, non-personal approaches of large corporations.

Ruth Anderson: Equal Partner and Hard-Nosed Controller

During the first fifteen years of Belcan, Ruth Anderson not only served as office manager and bookkeeper, but she was an equal partner and team member being involved in all decisions. Ruth remained in Cincinnati and

Ruth Anderson, office manager in the 1970s.

served as base. Her previous experience with bookkeeping in earlier jobs gave her the techniques needed to protect the home fort, Belcan, while Ralph was scouting the unknown for competitive advantage.

Even after normal accounting functions were assumed by others, Ruth remained at the business during the 1970s performing the control function of management. Her presence in the company helped police the activities of personnel.

Ralph was outgoing and trustful of everyone he dealt with. This managerial style encourages loyalty and teamwork; however, even in such a style, one needs a control system, no matter how simple. Ruth performed this function and identified the times individuals might be taking advantage of her trustful husband. Ruth provided this hard-nosed control for Belcan until the company became large enough to include a comptroller.

Ralph Anderson: Entrepreneurial Engineer

The period 1957 to 1961 was the leanest of Anderson's business life. For the first two of these years, he was sales manager at Ketco, the Schakel company that remained in Ohio. Anderson also found a few projects for Belcan.

His innovative spirit remained active. He obtained Patent #2,994,127 for Mo-Gard, a simple plastic device to protect the ankle and foot of a person using a power lawn mower. Mo-Gard merely strapped over the ankle, keeping the foot from sliding under the mower. Anderson had 10,000 of these devices produced and placed them directly into retail stores. Soon afterwards, however, he became concerned over the potential liability to him if someone was hurt using the safety device. He then recalled all items, destroying most of them. The patent produced no net income.

During his period of active engineering work, Anderson developed a reputation for testing stress in aircraft engines. Once he developed a system for testing an item, he found that he could make use of the same system when others sought his help. This experience gave him respectability as an engineer among prospective clients.

Anderson's skill in engineering opened opportunities for him during this difficult period of growth. Consistently, interviews with persons who hired Anderson at this time recall him as hard-working; a person who would deliver when he said he would. While other would make promises to deliver on a definite date, Anderson actually delivered. This meeting of schedules was important not only in this early period of Belcan but later in other projects, particularly in a multimillion dollar project for Procter & Gamble that was completely schedule driven.

GE engine mount at Lebanon, Ohio for Kett.

Anderson was fortunate during the first few years of Belcan to have made contacts with satisfied managers in previous activities. His work with General Electric in 1955–56 resulted in close association with Jack Hope, his boss at that time. Later, Jack Hope became an associate of Anderson's over several decades.

Anderson quickly became recognized as a competent engineer. But he believed that his most outstanding skill as an engineer was his sales ability. He continually stressed the need for technical people to focus on sales, that is, to secure clients and to cement their repeat business. Anderson believes that most engineers make poor salespersons; additionally, he believes that engineers make poor entrepreneurs since entrepreneurs must sell themselves as independent operators.

Allstates Design and Development Company, Inc.: Model for Belcan Corporation

With Kett, Anderson had become experienced as manager on a project having approximately ten engineers. After he left Ketco, Anderson became a contract engineer with Allstates, Inc., a company that hired engineers to serve clients of larger corporations on a temporary basis. In this type of firm, the engineer was the employee of the engineering firm, and paid a salary and fringe benefits by the firm but worked in the client's plant and was supervised by permanent employees of the client. The client paid Allstates a fee for the work performed by the contract engineer. It was around this concept that Anderson later formed Belcan.

Anderson focused on building Belcan's file of prospective engineers for temporary employment while he served as temporary manager as an employee of Allstates. At times, during the early period of Belcan, Anderson hired himself as a temporary employee.

Anderson made many contacts while at Allstates that were later important to Belcan. K. O. Johnson was one. Johnson was at Allison in the early 1960s when Anderson managed the group of contract engineers. Later, Johnson enjoyed a distinguished career at General Electric. In 1986, Johnson took early retirement from General Electric to work at Belcan in its Alliance with General Electric. (See *Profile: K. O. Johnson.*)

Karl Schakel and Kett helped Anderson develop as a project manager, but it was Allstates that demonstrated to Anderson the advantages of an engineering firm that concentrated on providing temporary employment for engineers. As an employee of Allstates, in 1962, Anderson worked on a number of interesting projects including stress analysis of the Re-entry shuttle for Aeronca at Lebanon, Ohio.

During the period from 1962–1964, Anderson managed 100 engineers in the Cole Building for a project at Allison Engines in Indianapolis. Homer Kallaher, a manager at Allstates, was Anderson's superior during this Allison project. Kallaher then left in 1967 to work for Wilde & Krause, but in 1969 he was hired by Anderson as Vice-President of Belcan. Anderson switched roles several times with individuals. He was first an employee then the employer of Jack Hope, Homer Kallaher, and K. O. Johnson.

One of the engineers who worked for Anderson at Allison in Indianapolis had a most interesting background. In May of 1993, Kenneth Rowe visited Anderson, his old boss, in Blue Ash, not having seen him for 15 years. Anderson called the author in Lexington and explained that he had omitted the name of an old friend from the 50 names which had been prepared in 1991. He said that the fellow, Ken Rowe, was from North Korea. Thinking that Ralph was joking, the author took the phone and asked what his name was. Ken Rowe then asked whether he should give his Korean name to which the answer was yes. He gave the name, No Kum-Sok, and told the following story.

Kum-Sok had been born June 10, 1932 in North Korea. During the Korean War he was a MiG-15 pilot, and on September 21, 1953 he became so unsettled under the Communist regime that he decided to leave. Kum-Sok was the pilot who first delivered a Russian fighter plane intact to the United States from North Korea, an incident that was a major news story at the time. After being protected in Okinawa and other places by the U.S. government for a decade, he returned to his major interest, jet engines, and turned up in Indianapolis in 1964 to be hired by Anderson. As of 1993, Rowe was teaching at the Embry-Riddle Aeronautical University in Daytona Beach, Florida.

In 1966, Anderson went to West Palm Beach, Florida to interview engineers who had been laid off from Pratt & Whitney. He selected one hundred engineers, most of whom he used in the Allison project. It was on this trip that Anderson became convinced that he could select a good employee if given ten minutes of time.

Gordon Bell refers to the "short-socks" test* as a basic factor in making a first impression in short interviews. Anderson makes decisions from his "gut feelings." It is through this feeling that he makes selections of persons in short interviews. It cannot be emphasized enough that a very important ingredient of a successful manager or entrepreneur is the ability to select personnel who fit the needs of the situation.

From 1962–1964 Anderson lived in Indianapolis during the week while working at the Cole Building in downtown Indianapolis; he returned on weekends. Ruth remained in Cincinnati and kept Belcan's offices open.

The first major project produced under Belcan's name was with 70 engineers for Allison in Indianapolis. This time, the location of the operation was in the Park-Fletcher complex near the major Allison plant located close to the airport. The project lasted from 1966 to 1968. The Park-Fletcher office was the first for Belcan outside of Cincinnati.

Haeco and the Giesel

Jack Hope, as Vice President of Cummins Engines, had been asked by Irwin Miller, its Chairman, for advice concerning a new type of engine that was

*A New England venture capitalist commented to Bell as to why a certain entrepreneur would not be funded by him, "because the President was wearing short socks." Bell names any specific element that affects the quick first impression as part of a "short socks test." C. Gordon Bell, HIGH-TECH VENTURES: THE GUIDE FOR ENTREPRENEURIAL SUCCESS, 1991 Reading, MA. Addison-Wesley Publishing Company. p. 16. Anderson appears to have no "short socks test" but refers consistently to his "gut feeling."

being made available by a Canadian group. Hope advised against the purchase, but started work on a new idea conceived by Robert D. Johnston who had been sent as a temporary employee by Anderson.

Hope referred the idea to an eminent evaluation team consisting of Dr. Lamont Eltinge (later to be President of Eaton and President of the Society of Automotive Engineers), Peter Schultz, President of Porsche, Stanley Jenkins, who had worked on the Nomad engine in England, and Dr. Roy Kamo. On October 28, 1968, the team gave its report.

Although the committee pointed out advantages of the giesel, Miller of Cummins was not interested. He explained that Cummins already had 50% of the truck engine market; the government would frown on a greater percentage. There was no payout for Cummins.

Hope left Cummins to give full time to promoting the new engine; he operated out of his family farm in Hillsboro, Ohio. Hope's work on the new engine was interrupted by a call from the White House asking him to be a Science Advisor for the President.

In the late 1970s, Anderson obtained a patent on this revolutionary engine, #3,498,053, which later became known as the giesel (turbo-charged diesel). The development of the engine would require millions of dollars. Several characteristics of the engine were attractive to the army, especially for use in tanks, since the engine required no water cooling. Belcan was successful in several proposals for development contracts with the U.S. Government. The Defense Department approved several hundred thousands of dollars in funding to Belcan.

A separate company was formed: Haeco, Inc. Much of the professional efforts of Belcan for the next six years were devoted to the giesel.

First Additions to Personnel and Expansion of Space at Belcan

After the completion of the Allison work in Indianapolis, Anderson was ready to expand Belcan's operations in Cincinnati. Belcan moved from the small office at 9505 Montgomery Road to the second floor space above a restaurant/bar at 9546 across Montgomery Road. This new office consisted of several rooms that could accommodate twenty to thirty employees with the required desks and drawing boards.

On April 1, 1968 John Kuprionis became the first permanent engineer besides Anderson in Belcan. At this time, Anderson was just completing the Allison job at Park-Fletcher in Indianapolis. Knowing the completion of the

Anderson as an entrepreneur in 1970.

project would mean the loss of 70 engineers, Kuprionis phoned Anderson asking whether he still wanted him since business would drop off precipitously. Anderson responded with a comment indicative of a true entrepreneur, "John come ahead right now . . . I'll need you more than ever; we will need to get out and build the business . . . I'll need your help!" (See *Profile: John Kuprionis.*)

With expenses increasing, Belcan expanded in size with billings for the first time exceeding $1,000,000, but profits declined. On October 6, 1969, Kuprionis recruited a contract engineer, Robert Smith, who became a permanent employee in 1975. In the 1990s he was the second oldest executive of Belcan in length of service. (See *Profile: Robert Smith.*)

Office Management: Ruth and Nita

On May 18, 1970, the bookkeeping staff was doubled in size by the addition of Nita Yoder. Nita's husband, Eugene, business manager for the Cincinnati *Post,* had died on February 9, 1970 of cancer. Since Nita had worked previously in bookkeeping, she interviewed with Ralph and Ruth for the job.

Decision making by Anderson might be specifically illustrated in the hiring of Nita Yoder. Nita reported that she was interviewed on a Friday afternoon for about an hour. Before she left the office, she was advised by the Andersons that they would like to check with their CPA for his professional opinion of her qualifications. Nita left the office. Within an hour, she received a call from the Andersons, telling her that they had talked over the matter and had decided that they should hire her without further checking. Anderson asked her to begin work on the next Monday. Nita, who lived in Mount Washington, could not drive an automobile at the time and asked her sons to drive her to work and to teach her to drive. Nita quickly proved to Anderson that she was a dependable person.

The operation of a temporary services office requires a system for billing clients and writing payroll checks. It needs offices only for files on engineers and clients and spaces for interviewing. The sales function involves developing relationships with clients who need temporary help. Until 1970, Ruth had held down the position of Office Manager with two administrative helpers for all records and filing for the company.

With the addition of Nita Yoder, Ruth and Nita developed a joint working relationship that served for the next five years. Each handled all types of business transactions such as billing, payroll, etc. The team worked together. On Mondays and Tuesdays they wrote checks to about 40 temporary contract engineers based on their time sheets; on Wednesdays and Thursdays they wrote invoices to about twenty companies who used Belcan engineers (these companies included General Electric, Babcock, Wilcox, Procter & Gamble, and Aeronca). On Fridays they cleaned up any loose ends and other bookkeeping chores. The entire set of books involved a peg board, and single-entry bookkeeping system by which the checks and invoices were written jointly using carbon paper.

In the first two years of the 1970s, billings doubled. At the end of each month, Nita would close the books. Before Anderson would ask for the total, he would scribble an amount on a piece of paper, then he would ask her for the exact amount. Nita remembered that he would have estimated an amount within a few dollars of the exact amount. This ability to handle revenue and expenses in his head repeatedly was valuable to Anderson; he would handle rounded numbers and leave the precision to Ruth and Nita.

During the late 1960s and early 1970s, the office had an account showing "back wages to Ralph Anderson." During this period, funding was tight and thus, Anderson would not draw the full amount, leaving that cash as working capital. Nita reported that, on the other hand, with employees, Ralph would round upward when figuring their pay—"he always gave the benefit of the doubt to the other fellow."

The tone of office management set by Ruth and Nita continued throughout the 1970s. Nita had a quiet manner, excellent for handling stressful situations. Since the office dealt with the records of the temporary engineers, the human relations of the company centered in this office. The observation made by John Kuprionis was that Nita was the perfect "PR" person. Her sense of humor carried to many small matters. Ruth reported that when she was trying to lose weight once, Nita tempted her with a hot fudge sundae and then snapped a picture when Ruth had completed the entire dish.

The quick choice of Nita as one of the first additions to the staff of Belcan is another example of the success achieved by Anderson's "gut feeling." She quickly fit into the young organization and helped build the optimistic view toward life held by Anderson. Once during an especially rough time for Belcan, she responded to a friend who showed concern about Anderson. Nita responded "Oh, don't worry about Mr. Anderson. I've seen him do his best work when his back is against the wall." Nita Yoder remained a loyal and dependable employee of Belcan until her death in 1991.

Turbine Power Systems, Inc.

In 1970, Robert Mendenhall, one of Anderson's friends before the formation of Belcan and who was then working for General Electric on turbine generating systems, came to Anderson with an idea; he and Anderson should form a company to produce portable turbine power generators. Anderson would handle the financial requirements and Mendenhall would work on production. Mendenhall had worked with a GE engine but he had in mind using the old Curtiss Wright J-65 engine and have it ready for sale within 90 days. Anderson responded that the idea was good and he felt that he could cover Mendenhall's salary for ninety days. The question was: who would buy these generators?

Anderson told Mendenhall that he had good contacts and relationships with General Motors' Allison Division in Indianapolis but their engine was much smaller than what Allison typically dealt with. Both went to see Anderson's contacts in Indianapolis where they were referred to the General Motors overseas division in New York. GM agreed that it would handle the sale of the portable generators with a $17 million contract for Iraq and Egypt; Anderson and Mendenhall would produce the engine.

In order to gain financing, Anderson went to his banker at Provident Bank of Cincinnati. He explained that he needed $100,000 for a down payment on a building to be secured with the contract that General Motors was willing to sign if he were ready to produce. Anderson had used Provident for Belcan since its start but generally only borrowed a little money at a time and paid it back promptly. Provident had commended Anderson for explaining opportunities of the engineering business to them, however, they hesitated to lend money based solely on a contract to purchase. Anderson, however, convinced the banker and secured funding for the building.

At this time, Anderson and Mendenhall incorporated Turbine Power Systems. Anderson gave Mendenhall 50 percent equity and a salary for his work. Seventeen turbine generators were built on trailers and delivered through GM's Overseas Division, with three going to Iraq and 14 to Egypt. This was Anderson's first venture into international trade. General Motors was interested in dealing with Turbine Power for its production for overseas sales because the company was small and could not short circuit products to customers, bypassing GM.

Anderson and Mendenhall tried to sell other generators but failed. The salary to Mendenhall lasted not 90 days, but four years with no significant extra income. Mendenhall had thanked Anderson for the arrangement, saying

Turbine Power Incorporated portable generators delivered to representatives of Iraq and Egypt.

that no one else would have given him a salary and 50 percent of the company with no financial contribution; however, at the end of four years, Mendenhall attempted to take over the company, literally locking Anderson out. Legal action ensued.

Anderson contacted Roger Penske who had close relationships in diesels with General Motors and who wanted to get into turbine engines. Penske bought Anderson's half of the company after the friction developed with Mendenhall. Mendenhall remained with the company for a year after the sale. Penske later sold the company after losing considerable money in producing units for South American countries.

Turbine Power Systems, Inc. represents an abortive venture by Anderson to expand into production through a separate company. It was an attempt to venture outside the scope of Anderson's distinctive competencies. It did not fit the strategic pattern for growth, i.e., temporary engineering services.

Turbine Power Systems, Inc. had no formal connection with Belcan. In fact, it took Anderson's time away from building Belcan and entangled Anderson with personal legal difficulties. Mendenhall continued to operate out of his home and remained in contact with Anderson but he never was an employee at Belcan.

Turbine Power Systems, Inc. illustrates how Anderson formed a company with his own financial backing even though he had limited personal resources. Although Belcan was 12 years old by 1970 and had contracts with firms like Allison, Ford, and others, the company had just reached slightly better than break–even status. Anderson had gradually convinced his banker that his type of engineering firm had a promising future.

Mendenhall and others looked to Anderson as one who would back them and would take risks that they would not take. He had the reputation of making a square deal and trusted his friends. At times, his trust cost him not only money, but time that could have been better used in alternative ways.

Also, Turbine Power, Inc. is yet another case in which Anderson capitalized on his superior work with Allison in Indianapolis. His reputation in General Motors could be translated into sales opportunities, even in overseas markets. He had already used his contacts at Allison for recruiting people to work for Belcan and in selling to clients.

Belcan in the Early 1970s

During the early 1970s Belcan depended on the technical services operations for its growth. During this period it expanded its base of large clients. It had good relations with General Motors and General Electric and now it began to develop work with Procter & Gamble. Later, this early growth in technical services served as a springboard for growth in full service engineering. By then, technical services remained important as the faithful cash cow.

Belcan's clients were affected by the economic cycles, and when their business was down, Belcan's profits would be down. In looking back, Anderson could identify a three- to four-year cycle in profits. He felt that this cycle indicated that his personnel tended to relax when times became better and thus create their own effort cycle in sales. Anderson was a salesman of engineering talent and new ideas. He felt that he could select people who could be their own self-starters and who would respond to a "pull" approach to motivation; yet, at times, he returned to the view in which he needed to "push" others to exceed. However, the gross billings continued an upward climb as can be seen in Figure 1.1.

The giesel continued to receive Anderson's attention in addition to Turbine Power Systems. For a time, John Kuprionis worked on the giesel. Hope and Johnston continued to provide some effort; however, the project needed funding up to $100,000,000. The Defense Department was the only source of funds for this development, but the grants were at the level of $100,000. Nevertheless, Anderson was intrigued by the efficiency of the engine. He devoted his personal attention and the efforts of Belcan toward its improvement primarily because of his technical interests and enjoyment of engineering design, testing and development. As Miller of Cummins said, "The technical aspects of the giesel were outstanding but the immediate returns were questionable."

The pattern developed by Anderson was to try out new ideas and to enter into them with different associates. The technical services operation tended to be too routine to keep his interest. He thrived on assuming new risks and depending on his cash cow to carry him along as he continued to do what he really wanted to do.

Real Estate as an Investment in the Future

In 1974 Anderson purchased a five-acre lot on Deep Run Road in Indian Hill on which he built a house for his family. Whenever he could, he swung a financial deal to buy a building or land. By the 1970s he had moved from renting property, which had been necessary in the 1950s, to purchasing real estate. From the late 1960s, investment in real estate made a major contribution to Anderson's acquisition of wealth. During this period,

Anderson purchased farm land in Mercer County, Kentucky, as a means of building up his savings. Inflation during these decades made these investments especially wise.

Banks and Cash Flow

In the seventies, cash flow was handled by routine borrowing of working capital from Cincinnati banks. During the 15 years of rapid growth, Anderson continually had to fight for established city bankers' support. For most of the early period, Anderson used the Provident Bank as his major source of working capital. He developed a close working relationship with the bank and kept them informed about the unique need of his type of business. However, just when he had developed a smooth relationship with the bank, it was purchased by another bank. After that time, Anderson continually had to devote time to educating his banks about his business. Some banks felt that he was too small to worry with; some felt that he took risks that were too great for the bank. The cyclical nature of the business raised questions with other banks, especially when Belcan had a loss in a year or two; some banks felt that Anderson was using the funds for purposes other than working capital.

During the following 20 years, Belcan changed its major bank six times, averaging about three years per bank. Conservative banks found different things to worry them. The fact that Belcan's bottom line showed profits most years made it possible for Anderson to finance operations with short term borrowing but the cyclical nature of the business caused bankers to be concerned.

Operations in the Mid-Seventies

We have seen that Belcan depended upon unique techniques for handling technical services for major companies. Belcan grew as the larger companies became aware of the advantages to them. A national association developed among "design" firms with over 500 firms ranking members according to size. Many of these firms specialized in one type of engineering: some concentrated on water supply; others on power, others on manufacturing, building, sewer waste, transportation, or hazardous waste. By the 1980s Belcan was ranked 26th in size in the nation and showed a major percentage of effort in manufacturing. The rank changed from year to year but Belcan remained in the top 100 of the nation's engineering firms.

Testing jet engine controls.

A brief caveat should be stated about interpreting sizes in the specialized fields of engineering. Belcan has two types of engineers listed as its employees:

- Those **contract engineers** who are hired by Belcan to work for Belcan's clients at the clients' plants. Occasionally, these employees will
  work on site at Anderson Way on a specific project. Some engineers
  prefer the flexibility and diversity of this arrangement and work in such
  a capacity for many years.
- Those **direct employees,** hired permanently by Belcan who work at
  Belcan's own locations.

The rate of growth in number of employees and the expansion of the
mission remained stable from 1968 to 1976. Until 1976, Belcan appeared
to be very small with offices on the second floor of a restaurant/bar on
Montgomery Road in Cincinnati. We shall see in the next chapter that by
1980, Belcan operated as a middle-sized firm with modern accommoda-
tions as a national firm. The story of this take off stage is the story of its
growth in "in-house engineering."

One final note regarding size. Most success in business is measured by growth and by being a large firm. This is especially true in engineering services whose product is satisfied customers. When each customer has unique needs, increasing size has advantages but it also carries with it increased difficulties. One of the main advantages of a small firm is the personal touch of the management and how the flexibility of strategies and policies to mesh with those of the clients. Anderson's management approach was especially suitable to the smaller firm. During the take-off stage of the 1980s, some of the difficulties of growth and Anderson's management style began to appear.

PROFILE OF K. O. JOHNSON

The career of Kenneth O. Johnson, generally known as K. O., has similarities with that of Ralph Anderson's except that K. O. had a distinguished career in management and Ralph became an entrepreneur. They are close in age, both men first worked for Curtiss Wright, and both were in the Air Corps, Ralph as an engineer, and K. O. as a P 47 pilot. Employment with GM's Allison was where the two met in the early 1960s; GE turbine engines were important to both, K. O. as a designer and Ralph as a contractor.

K. O. was born August 31, 1922. He graduated from Purdue with a BS in Aeronautical Engineering. After K. O. joined GE in 1966 he introduced and patented the UDF Engine, which included big improvements in fuel consumption. In 1969, he had been recognized for management of the F100/400 Core engine and continuing through concept and early development of the F101/CFM core. Many awards and recognitions came his way for work on engines, including the LM2500 for the Navy and for more than six industrial engines. He also received an award on March 5, 1987 for a technical brief on the Unducted Engine.

K. O. took early retirement from GE but immediately came to Belcan and continued to work with Jim Young in the GE Alliance. Even for an individual with such a distinguished career with a large company, Belcan offered a working environment motivating to a person past normal retirement.

PROFILE OF JOHN KUPRIONIS

In 1968, John Kuprionis became the first permanent professional employee of Belcan. Kuprionis became Vice President early and has stayed in Services Group throughout his tenure. The fact that technical services was Belcan's initial mission and remained its most stable area and the fact that Kuprionis could be depended on to handle its management gave Anderson security as he ventured into more risky avenues.

John Kuprionis was born on January 14, 1938 in Vilnius, Lithuania. His family later moved to Kaunaf and, during the Nazi occupation, to a small rural village. As the Russian army advanced into the country, many of his countrymen moved west to avoid Communist occupation. Within hours, his family escaped from the Russian army and moved to Gras, Austria in order to be in the allied zone at the end of the war. He attended school for one year in Augsburg.

His family moved to Sioux City, Iowa, in 1946, where John's aunt lived and sponsored the family in immigration. Although his father spoke five languages as a professor in Lithuania, he could not speak English. Thus, he had to work in a meatpacking house.

In Lithuania, John's father was not only Professor of Forestry at the University, but was a ranking official in the Lithuanian government. For this reason, when the Russians entered the area, soldiers sought his father by name. The family got away by train six hours before the arrival of the Communist army. A relative who was in railroading had stayed behind to evacuate John's family was never heard from again.

From Sioux City, John, his father, mother, and two younger sisters, moved to Chicago, where John's father was a manual laborer in the park system to get him closer to his specialized field. Later, John's father got his Master's degree in forestry at Michigan State University in Lansing, Michigan. Finding no academic position after learning English, he became manager of a landscape company in Dallas, Texas. Finally, his father became professor of Forestry at Louisiana Technical University, which John attended.

After obtaining a degree in mechanical engineering from Louisiana Tech in 1958, John Kuprionis served two years in the U.S. Army as a draftee. He joined Borg Warner to work on a subcontract for a nuclear powered submarine for the navy. But John wanted broader experiences in engineering, and he chose to join Allstates as a contract engineer in 1964. It is from this job that Kuprionis became a permanent employee of Belcan in 1968.

As Vice President of Belcan Services Group in the 1990s, John, with his engineering background, teamed with McCaw's legal education and with Donnelly's business focus to provide a well-balanced expertise to Belcan Services Group.

PROFILE OF ROBERT SMITH

In October, 1969, when Bob Smith arrived at Belcan with its offices still on Montgomery Road, he did not know that he was getting more than a temporary job. He had been a contract engineer working for several aircraft firms in the West, and living in Reno, Nevada, but after meeting Sandy O'Leary, who was already working at Belcan, he married her and moved to Cincinnati.

As a contract engineer with Belcan from October 1969, he worked at Aeronca, Procter & Gamble and at General Electric on the F101 and F401. During this time he worked part time on his engineering degree at the University of Cincinnati and received it in 1975.

Bob Smith was born on an upstate New York farm in 1945 and discussed farming with Anderson in the 1980s as Anderson began to build up his holdings in Mercer County. During most of his long tenure with Belcan, Smith filled in many different positions and helped provide flexibility to Belcan during its rapid growth in the 1980s. At this time he became interested in the Deming approach and supported the development of Belcan's Continuous Improvement Program. During the middle to late 1980s, Smith served as Vice President.

After the reorganization of 1988–1989, Smith joined Jim Young in support of the General Electric Aircraft Alliance and continued to serve loyally as utility team player.

Full Service Engineering—
Stage of Rapid Growth, 1976–1985

"Employees who enjoy their work will be the most productive."
RALPH G. ANDERSON

The decade from 1976–1985 was the "take off" stage for Belcan. This tremendous growth resulted from the decision to move into "in-house" engineering. Technical services, the dependable cash cow, provided stability but did not offer the challenge of a complete engineering company. To offer engineering services on a project basis required additional capital for expansion in plant capacity whereas expansion of temporary services required merely the addition of several offices with filing cabinets and a computer system.

Industrial Environment

The rapid changes in the industrial environment in the 1970s and 1980s set the stage for entrepreneurs to fill new niches. Maintenance of large engineering departments by major industrial firms, such as General Motors, General Electric, and Procter & Gamble became increasingly expensive primarily because of the enlarged packages of benefits (hospitalization, pensions, etc.) caused by new legislation, union pressures, and social expectations.

One response for containing cost was to shift the expense to a specialist who could better utilize professional personnel and handle the "fringe benefits." Belcan was one of the firms that seized the opportunity to serve these large corporations. The expansion did involve greater risks because greater capital was required and the action carried with it "full service liability." Belcan had the advantage of being more flexible in the use of professional engineers. When concentrating in engineering services, Belcan could keep a larger variety of types of engineers and perform services for its clients with greater expertise. Professionals could be moved from one project to the other and thus avoid the idle time that some large firms experienced. In short, an "in-house" operation could do a better job at less cost for its clients than their clients could do for themselves.

The Cincinnati area has long been an industrial center with heavy industry, machine tools, and major plants of the national firms. Furthermore, the area from Dayton to Evendale is a major center for the aircraft industry. The Ohio River Valley is centrally located for access to the Midwest industrial market. The area is home to Ruth Anderson and it is the nearest industrial area to Ralph Anderson's home in Mercer County, Kentucky. Thus, it is not surprising to find Belcan expanding in the area of Cincinnati at Blue Ash, Ohio.

Technical Services and the Move into "In-House" Projects

Anderson's choice of first offering temporary engineering services was favorable to a "new kid on the block." It required small financial backing. Its success was dependent on the respect and honesty of an employer who would work for the benefit of his employees even if the employees were not physically working with the employer. Furthermore, the opportunity for obtaining excellent personnel for the permanent growth of Belcan through first hiring them as temporary employees for another firm enabled Belcan to staff its own company with people who had first been tried as temporary employees. Several Belcan executives who were still with Belcan in the early 1990s were first hired as temporary workers. In fact, in the early days, Anderson was also able to give himself a "temporary job" when opportunities for permanent work were not available.

"In house" refers to the location of projects of clients within the engineering service company, as opposed to hiring temporary engineers to work in the clients' own plant. Later, the concept became known as full service engineering because the engineering company took on greater responsibility and liability for all work within a designated project. The decision created a greater need for management to develop rapidly to satisfy the needs for the expanded operations. Opportunities were great, but they carried increased risks; just the combination that Ralph Anderson thrives on.

Until the 1970s, large industrial firms had several engineering departments and divisions within the parent organization in which they performed all of the engineering functions themselves. They did make use of engineering consultants and at times they needed the help from temporarily hired engineers, but the focus of all the engineering work was within the parent organization.

With changes in the competitive climate, especially from the Japanese who operated with different ideas, and the increased costs of fringe benefits to employees, the large firms became interested in contracting out much of the engineering to outside specialty firms. As a small firm with the flexibility to move quickly, Belcan was able to adjust to changes and to grow rapidly to meet new challenges. This growth was breathtaking.

Jim Young and the Giesel

Anderson knew Jim Young at Allison in Indianapolis. In the spring of 1974, Anderson asked Young to have lunch with him and to check out the giesel during an afternoon. Anderson explained that he had a contract with the United States government for development of the giesel, but the money was contingent upon Belcan's having the giesel operating by July 1, 1974. Since the government funding was drying up for the space programs on which General Electric, his employer, was dependent, Young saw the need to seek new challenges. Young's ability was in design and his assessment of the giesel was that it had significant design flaws. (See *Profile: Jim Young.*)

On May 1, 1974, Jim Young joined Belcan with the immediate objective getting the giesel running by the government's deadline of July 1, 1974. Bob Johnston and Jack Hope had been trying to get the giesel to run since February. At this time, Hope was in Washington D.C. as Science Advisor to the President and later, moved his files to his farm in Hillsboro, Ohio. Disagreement with Johnston and the unavailability of Hope left the further work on the giesel in the 1970s to Young and Anderson.

The second floor at 9544 Montgomery Road was becoming quite overcrowded. Lloyd Sheeran, Ralph and Ruth Anderson, Kuprionis, Smith, and Young were the chief permanent staff. Space would soon become a problem even if there were no changes in strategies. In 1974, Pat Wagonfield joined Belcan coming from Procter & Gamble. (See *Profile: Patrick Wagonfield.*)

Young was successful in solving the design problems with the giesel and got it running by the deadline of July 1, 1974. The government then approved the additional funds for development and Young stayed with the giesel project until all work was stopped after 1977. On this date, the Army demanded that the work be on a cost sharing contract. Such a contract would have required additional funding by Belcan—it could not be obtained.

The Deerfield Building in Deer Park

Young agreed that Belcan should move into "in house" work. If Belcan were to do this, additional space would be needed. At the time Belcan had no in-house work other than the giesel.

At this time, Procter & Gamble made the decision to outsource its engineering and obtain firms to do it on a project basis. Belcan had a number of engineers working on technical service contracts in P&G's facilities. If P&G eliminated engineering within its plant, a number of Belcan's contract engineers would have no place to work and thus, lose their jobs. If Belcan moved into in-house work, those engineers could be hired by Belcan on a permanent basis.

Anderson seized this opportunity and sought Procter & Gamble project contracts. In fact, Anderson hired 50 or 60 engineers from P&G before he had sufficient space to house them or sufficient work to justify their hiring. It was during this period that Anderson developed a strategy to always build buildings ahead of the demand; the availability of the space gave him an extra argument for obtaining the contract.

By early 1982 Belcan not only attracted P&G's engineers but also experienced managers who could recognize the changes in the industrial structure and move to smaller engineering specialists before being laid off. Dan Swanson, knowing the quality of Belcan's work and the internal operations at P&G, decided that it was time for him to make the move. (See *Profile: Dan Swanson.*)

The decision to go "in-house" and to secure the P&G contract resulted in the construction of a building at 11083 Deerfield Road. Originally, a 6,000 sq. ft. building was planned, but almost immediately Anderson doubled it to 12,000 sq. ft. because of his belief of first having space available. By the time the building was completed in April of 1976, the entire 12,000 sq. ft. was used and, in fact, it was necessary to lease 10,000 additional square feet nearby in a building known as the Junior Achievement Building.

Opportunities were developing so rapidly that during the completion of the Deerfield Building, land was purchased for a much larger building near the Blue Ash Airport, later to be called Anderson Way. See Table 5–1 for a summary of Expansion of Belcan's physical plant.

Thus, full service engineering became a major thrust of Belcan along side Belcan Technical Services. Jim Young became the director of Engineering Services. Belcan could expand more rapidly in the Deerfield Building

Table 5–1

SUMMARY OF BELCAN'S PHYSICAL PLANT				
DATE	**ADDRESS**	**NUMBER OF EMPLOYEES**	**SQUARE FEET**	**REMARKS**
1958	9505 Montgomery Road	2	One Room	Ruth and Ralph only
1966–1976	9546 Montgomery Road	20	3,000 sq. ft.	
April 1976–1977	11086 Deerfield Road	100	12,000 sq. ft.	Lease Junior Achievement Building
Jan. 1980	10200 Anderson Way	300	52,000 sq. ft.	Blue Ash Headquarters
Nov. 1983	7725 East Kemper Road	Variable	45,000 sq. ft.	
1985	R.E.O.P. Pittsburgh, PA.	200	55,000 sq. ft.	Du Pont Alliance
1986	10200 Anderson Way	300	52,000 sq. ft.	
1986	BGP, Northland Blvd., Cincinnati	450	54,000 sq. ft.	Addition to Headquarters
1987	S.E.E.D. Solon, Ohio	200	50,000 sq. ft.	
1987	DCAT, GM, Dayton, Ohio System Product Dev. Center	35	0	Delco, GM Alliance
1987	Lodge and Shipley 1328 Elam St., Cincinnati, Ohio	0	200,000+ 14 acres	Plant sold
1989	Eli Lilly, Lafayette, Ind.	Variable	Variable	Lilly Alliance
1989	11591 Goldcoast Drive	40	10,000 sq. ft.	Technical Services Headquarters and Temporary Services
1992	McGraw Engineering 340 17th Street, Ashland, Ky.	75	10,000 sq. ft.	
1992	CSA Merchantil Plaza Hato Rey, Puerto Rico	65	5,000 sq. ft.	

but Anderson started construction of a much larger building on the land in Blue Ash. In January of 1980, Belcan moved to Anderson Way with its 52,000 square feet of space.

The larger company did not have an accounting department until it moved into the new building. Ruth Anderson and Nita Yoder had efficiently handled the bookkeeping (invoicing and payroll) but the other office functions were at a minimum. In June of 1980, Michael McCaw, who had an accounting and law degree, came into the company after practicing law in Cors & Bassett, a Cincinnati law firm, for a year. Candace, Anderson's daughter, joined her husband, Mike McCaw, in the accounting operations. (See *Profile: Mike McCaw*)

Revenue continued to increase in the early 1980s in spite of the national recession. With increased costs of the new building and the extra operating costs of larger companies increased overhead, earnings in the early 1980s were low or negative. Anderson responded to these weaker years by expanding the company personnel and by increased emphasis on sales.

Two executives were recruited in the early 1980s: William Thomas and Lane Donnelly. Thomas had worked for a similar type of firm and exhibited a strong salesmanship image. Donnelly had been a teacher after graduating from a business school and had over ten years experience with personnel activities of similar engineering firms. (See *Profile: Lane Donnelly.*)

William Thomas was brought into the company from Butler Services Group on July 11, 1983. A Belcan Vice President had met Thomas and was impressed with his views of salesmanship. Associates remember him as an exciting person. Thomas was made vice president and stationed in Belcan's new office in Atlanta. The office failed to reach its expected target in a reasonable time and Thomas was released on January 1, 1986.

On November 14, 1983, Lane Donnelly was recruited to be Vice President, National Sales, Belcan Services. From this date, Donnelly established a firm record in Technical Services and later became President of the separate corporation devoted to temporary services.

Anderson continued to expand building space ahead of actual work available. Work sufficient to fill this space seemed to always develop. This growth reassured him that his strategy was an important sales element. It was not until 1987 that rapid expansion of space resulted in costly idle capacity.

A major change in technology used by engineering firms took place at the time Belcan moved into its new building on Anderson Way. In 1979, Belcan purchased its first Computer-Aided-Design (CAD) system to perform the drafting work for a major client. During the first half of the

1980s, four mini-computer based systems were added, as well as two Calma CAD systems on Data General mini-computers with 15 work stations, one Calma CAD system with six workstations, and an Integraph CAD system with six workstations.

Up to this time all work had been performed using conventional drawing boards, hand calculations, slide rules, and, at times, portable calculators. By April, 1984, Belcan required a computer specialist. Thomas J. Le Saint was recruited from Procter & Gamble as CAD Section Manager. (See *Profile: Thomas Le Saint*.)

Procter & Gamble: Pampers Project 1984–1985

Procter & Gamble had perfected a disposable diaper in 1974 and had used several brand names, including Luvs for female babies and Pampers for male babies. However, Kimberly Clark had quickly entered the business and had successfully grabbed a share of the market for their Huggies. Procter & Gamble, known for its aggressive competitive behavior, immediately responded with a redesign of its products that would require the design of new machines in all of its plants. Competitive pressure dictated that the new machines be designed in minimum time.

The first stages for obtaining the huge design project began with the allocation of 5,000 square feet for a model shop in a new building on Kemper Road. Quickly, the space was expanded to 13,000 square feet and Belcan became committed to the largest single project in its history.

There was no time to sequence the design steps in normal order. Initial drawings would take too long; therefore, the project moved to the construction of wooden models for the 25 machines (lines) to be set up in four Procter & Gamble plants. Belcan would then contract out for the production of the metal parts as drawings were produced.

This process is a classic example of a "crash" program, that is, performing several stages simultaneously instead of in sequence. The success of the project depended upon rapid response between the trials of the wooden models and the revised orders for the metal parts. As the models indicated needed changes, the orders were issued for changes. Even though this process resulted in extra costs, the entire project was schedule driven. Time was of the essence—target dates were to be met if the project was to be considered successful.

The Pampers project required competent engineers who would not hesitate to make changes in the routine patterns learned in school. Additionally, management required a rapid communications system and cooperation among professional personnel. The flexibility that was a major characteristic of Ralph Anderson and his small, young firm was just the type that could show the "big boys" how to operate a tight ship with numerous components.

At one time, the strain on space became so great that a new 45,000 square foot building planned for operations on Kemper was used solely for storage of poly bags, part of the packaging of Pampers. Tight control over the subcontractors for the metal parts at one time demanded that Belcan take over the operation of the subcontractor in order to meet the tight schedule.

During the period from 1984–1985, the Pampers project required a major effort by most of Belcan's employees. Finally, the company hit every time target that had been set; the machines had been set up in the P&G plants enabling it to regain a major portion of the disposable diaper market. During the peak growth caused by the effort on Pampers, three most valuable people were recruited. (See *Profiles of Dennis Evans, Isaac Gilliam, and Jane York.*)

During this rapid expansion, Jim Young located a chemical engineer who was available after release from a small, local engineering firm. John R. Messick was hired to expand the business in process engineering after the Pampers project was completed. (See *Profile: John Messick.*)

Diversification

In 1978 Anderson laid the foundation for expansion into construction with the purchase of Wilde & Krouse, a firm with 12 architects. It was reasoned that designing buildings would increase the full service capabilities of Belcan. This later served as base for a subsidiary devoted to construction of industrial buildings. Ed Holland led in the shift into the construction of industrial plants, including refineries and continuous process units. The problem was that new managers, some with only home building experience, had no experience in construction of industrial plants. However, by mid-1986 Holland had opened three new construction offices: one in Pittsburgh, managed by Gordon Kidd, one in Cincinnati, managed by John Denise, one in St. Louis, handled by Holland. The pressure to grow resulted in obtaining large contracts for which the officers were not experienced or prepared. Within less than two years, the construction venture had proven to be a major mistake. Senior Belcan managers gave attention when other major problems required their attention. In short, Belcan not only had bit off more than it could chew, it had entered activities beyond its competence. As a much larger firm with more experiences, Belcan was thinking of the advantages of the strategy of design/build.

During the decade of the 1980s, Belcan acquired several other firms. See Table 6–1. In 1985, Belcan purchased Multicon, Inc., which had been founded by Joseph E. Campbell and Robert Flisik in 1982. Multicon was a high technology firm that used solid state TV for quality control and robots for applying industrial diamonds. Like other entrepreneurs, Campbell stayed briefly with Belcan but later moved out of the company. Multicon, with its high-tech image, was very enticing. Again it did not live up to its reputation and within four years, it had disappeared from Belcan.

Organizational Changes

Anderson devoted much of his time to acquiring companies. Thus, he needed executives and consultants to help him select new managers to operate the larger and diverse company. The Pampers project took a great amount of management attention. By 1985, several fundamental developments were occurring simultaneously. Growth almost literally caused an explosion of changes. The Pampers project was so large relative to other projects at the time that Belcan was forced to concentrate most of its attention to this major client. However, the client was so large and powerful that it could have a major impact on even the internal operations of Belcan. P&G was aware that the Pampers contract was the most important project held by Belcan and used its position to seek to further enhance it relations with Belcan.

Although Belcan had met all schedules and had performed in an outstanding manner, P&G tended to view Belcan's engineering, not as a separate company, but as a part of P&G. With this view, the larger company attempted to dictate the personnel with whom they would deal. Thus even though Pampers represented the "take off" stage for Belcan, it simultaneously laid the foundation for a major crisis.

Jim Young had made a major effort to build an "in-house" capability, but after the huge Pampers contract with Procter & Gamble, the client raised questions about the continuation of the contract. Jim Young was reassigned to duties suitable to his technical expertise and initiated new partnering relationships and alliances, which will be discussed later.

Anderson had made it a practice to select executive personnel with his "gut feeling" and with personal knowledge of their capabilities. Now, for the first time in the history of the company, it was necessary to recruit new personnel quickly through newspaper advertising and through the use of headhunters.

A professional recruiting company, Keith Baldwin, was given the request for three top executives: a human relations director, a general sales manager, and a controller. Until the 1980s, Anderson had depended on his tried and true method of personnel evaluation.

Gary Bates was hired on August 1, 1985 for national sales through recommendation by a consultant from New Jersey. In November of 1985, Bates was made Senior Vice President, General Manager of the Engineering Division. The period of 1985 can be summarized as a transitional year and one of great confusion. Preceding it was the period of "take off", rapid growth and increased profits. 1986 was the year in which past success encouraged an ill-advised expansion and change in basic strategy. On August 27, 1986, Gary Bates, after one year in the position of president, left Belcan to take a position with BGP, a related partnership to be discussed later.

In less than a decade, Belcan had grown from a company with total revenue of $5 million to a company of $80 million. It is not surprising that it had the usual growing pains associated with a change from techniques suited for a small company to those of a middle-sized company. This stage is typically a dangerous one for entrepreneurs. The need for more attention to management and organization is usually evident. We shall discuss in the next chapter the financial crisis and the consolidation that resulted from the financial strains of growth.

PROFILE OF LANE DONNELLY

L Lane Donnelly, a non-engineer, but with ten years of experience on the Board of the National Technical Services Association (NTSA) became Vice President of National Sales of Belcan on November 1983. This was the beginning of Belcan's rapid growth in "in-house" engineering. Donnelly and John Kuprionis concentrated on Belcan Technical Services and were anchors for Belcan's "cash cow" during the next decade.

Lane Donnelly was born on June 4, 1937 in Philadelphia, Pennsylvania. He moved to Camden, New Jersey where he graduated from Audubon High School, where he played baseball. Because Lane's high school coach knew the coach at Murray State College (now University) in Murray, Kentucky, Lane chose to attend college at Murray where he obtained a degree in business administration in 1961 while playing baseball. After graduation, Lane returned to Philadelphia where he joined an army reserve unit and became a part-time substitute high school history teacher.

Later, after returning from active duty, Lane was married to Betty Jean Ernest and became a full time sixth grade teacher in New Jersey. After teaching for three years and serving as assistant principal for the last year, he took a summer job in New York with his father-in-law manufacturing leaf springs for automobiles.

Influenced by his father-in-law, who believed that he would never make enough money teaching to provide Betty Jean with an acceptable lifestyle, Lane joined Allstates Design and Development Company in 1965 as their personnel specialist. He became established with contract engineers, which prepared him for his later move to Belcan.

Donnelly became Corporate Personnel Manager with Allstates, which at the time had about 1,000 employees, and also worked in marketing and client services. He moved to Cincinnati in 1978 as General Manager, South Central Region of Allstates. In October 1982, he became Vice President, Atlantic Services Group (of contract engineers), where he supervised specialty services to the electric industry. After his move to Belcan, he became President of Belcan Services Group in 1986 and remained the anchor in technical services for Belcan during its turbulent years.

PROFILE OF DENNIS EVANS

 Denny Evans began a lifetime interest in teaching at a Kentucky college. Later, however, he obtained an engineering degree and a position with Procter & Gamble. After joining Belcan in 1983, he served as project engineer for several years, but soon after adopting the ideas of Total Quality Management he returned to his original orientation as a teacher—this time as Belcan's educator in Total Quality Leadership.

Dennis Evans was one of twin boys born on June 28, 1944 in Fairfax, Ohio. He attended a parochial school for eight years in Madisonville, Ohio. At Marymont High School, Dennis was an athlete who qualified for college athletic scholarships but because of a broken leg and the lack of financial resources he had to work for a small hardware business for several years.

Deciding he needed additional education, Dennis became the first member of his family (his father was a milkman) to attend college, entering Morehead State University in Kentucky with the goal of becoming a teacher. After a year at Morehead, because of the prospect of low pay in teaching, he switched to engineering, entering the Ohio College of Applied Sciences (later the University of Cincinnati). Through their co-op program, he worked at Procter & Gamble while taking courses for the associate degree in mechanical engineering, completing it in 1966.

On November 19, 1966, after graduation, Dennis married. Evans continued to work in a variety of departments of Procter & Gamble for the next 16 years, first in paper products, then in process engineering (liquid detergents), and finally in the international division. Taking advantage of P&G's bidding systems for jobs, Evans qualified for a job in starting up a new plant in Madrid, Spain, in 1975. He was accompanied to Spain by his wife and family. The experience of living in a different culture expanded the horizons of the southern Ohio native. After returning from a year in Spain, he accepted a year in Caracas, Venezuela, after which he remained in P&G's international division, traveling extensively to South American countries in helping build seven new plants.

After P&G scaled down its rapid international growth in 1983, Evans accepted the "separation package" from P&G. Having been the client of Belcan for almost a decade, he knew about the small engineering company with which he had had good experience. He joined Belcan as a permanent employee, not necessarily as a career choice but because it offered immediate employment. However, in the new company he found greater freedom in the unstructured environment of Belcan and began to have fun in the jobs because they offered chances for creative activities. Evans moved to project manager and then to program manager (director), responsible for all projects with a given client.

In 1985 Evans began to concentrate on building a quality emphasis into all projects and programs. After the arrival of J. J. Suarez, several managers began to develop a Total Quality Management program. First, they developed Continuous Process Improvement (CIP) and later Total Quality Leadership (TQL). By the 1990s with strong support of President Suarez, Denny returned to teaching. This time his students were first Belcan employees, then Belcan clients as "billable services," then university engineering students, and finally, by 1993, Evans and Suarez were internationally known quality experts.

In 1993 Evans was made Vice President, a member of the Lead Team.

PROFILE OF ISAAC GILLIAM

 When Isaac was a first grader, he visited the house of the only architect in the Tennessee valley where he lived. From this moment on, Isaac knew that architecture would be his life work. Encouraged by family and teachers, he became the only member of his family to attend college, taking advantage of the University of Cincinnati's cooperative program. He graduated in 1969.

Gilliam was born on a 300-acre hill farm near Kingsport, Tennessee in 1944. His father worked for Eastman Kodak, farmed the land, and worked in his woodworking shop at home. Isaac began as a boy to save the earnings from cracking walnuts, tending his own few cattle and other odd jobs to make it possible to attend college. While in college he waited on tables, illustrated a text book for a professor, and continued to support himself through the six year college program. He married in 1967 while in school and graduated in 1969. From 1969 to 1972 he served in the Corps of Engineers during the Vietnam War.

After he was discharged from the army, Isaac took a position with an industrial architectural firm working on buildings in the dairy industry. In September of 1984, Isaac interviewed with Bob Smith and Roger Shutt and was hired immediately. Early in 1985, he met Anderson, who asked him to help with the renovation of the house on the Mercer Farm. Together with the Wilson brothers of Harrodsburg as contractors, Isaac supervised the remodeling of Walnut Hall and Wildwood.

Gilliam worked closely with Anderson in planning Belcan's facilities on Anderson Way, Kemper Road, and the buildings for the new Alliances in other cities. By 1988, Gilliam worked as a project manager with Mead Paper, expanding Belcan's work with that company. In 1990, Isaac became program director for the pulp and paper industry, he was one of five program directors of the company. In 1993, he became Vice President and a member of the Lead Team of Belcan Engineering.

PROFILE OF THOMAS JOSEPH LE SAINT

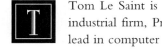 Tom Le Saint is a native of Cincinnati, an alumnus of its leading industrial firm, Procter & Gamble, and one who quickly took the lead in computer design. Tom became a critical resource of Belcan as it continued to seek state of the art technology in engineering design.

Le Saint was born on August 16, 1950 and grew up on the west side of Cincinnati, attending LaSalle High School. In August of 1971, he obtained the A.S. degree in Mechanical Engineering Technology from the University of Cincinnati, while working in the co-op program with Procter & Gamble. Le Saint began work full time as Associate Designer. For the next 12 years he moved up in Procter & Gamble's Design Engineering Division and was involved in many aspects of the paper converting technology, including the design, installation and startup of several new products. Some of these included Puffs facial tissue, Bounce fabric softener, and Certain toilet tissue products. He was also involved with numerous improvement projects to an existing array of products such as Bounty kitchen towels, White Cloud and Charmin toilet tissue, Pampers diapers, and regular Puffs facial tissues.

In 1979 Tom was trained as one of the first Calma–CAD mechanical designers, and in April of 1980 was awarded the honor and position of "Advanced Technical Designer" in the Design Engineering Division. For the next four years, Tom played a key role in the Computer Technology group, writing many utility programs to maximize the efficiencies of the CAD technology in the mechanical design and engineering applications.

In May of 1983, Le Saint was hired by Belcan as the CAD Software Development Manager. Arriving with Belcan just as the company was taking off in growth activities, Le Saint moved up in the organization as the chief computer specialist. In January of 1990 he became Director of Information and Computer Services with an increased scope of activities that now included not only engineering computing needs but those of business financials and project tracking as well.

Tom is married to Ann (Richards) Le Saint, and they have four children.

PROFILE OF MICHAEL E. McCAW

 Mike McCaw was born on April 10, 1954 near Blue Ash, Ohio. He attended schools in the area and met Candace Anderson, Ralph and Ruth Anderson's only child, in 1970 when Candace was a freshman and Mike was a junior in high school.

Mike and Candace were high school sweethearts during the early 1970s when Ralph and Ruth were operating a small engineering firm in a second floor walk-up above a bar on Montgomery Road. Mike's father owned a successful coin-operated machines business in the Cincinnati area.

Mike entered business school at Ohio State University in Columbus and Candace remained in Blue Ash for her final two years in high school. Then she, too, entered Ohio State. Soon, Candace transferred to a one year program in a school for dental technicians. This way, they both completed school at the same time, setting the foundation for their marriage.

Mike and Candace were married on August 7, 1976 after Mike had completed his bachelor's degree in accounting at State and after Candace had completed the requirements for dental technician. Both left for California where Mike entered Santa Clara Law School and Candace worked to supplement Mike's scholarships at the private school. The couple returned to the Cincinnati area and Candace went to work for Belcan to help her mother in the accounting department. Mike joined a law firm downtown.

In 1980, the accounting department for Belcan was essentially Ruth and Nita, with clerical help and an outside accounting firm to provide advice.

The New (1980) building, 10200 Anderson Way.

After Mike became bored with the routine practice of a large law firm, he recognized that his accounting degree from Ohio State gave him an excellent basis to help his father-in-law build a professional accounting department. After spending one year recruiting engineers with John Kuprionis in Technical Services, McCaw moved to recruiting accountants.

As Belcan grew rapidly, Mike saw that the company needed a more professionally trained accountant. John Reed, a CPA from the public accounting firm which had handled the Belcan account for several years, was hired. Mike subsequently added two or three people in accounting, eventually building the department up to fifteen people.

Mike had always been fanatical about golfing. The year was 1986 and he felt the time was ripe for him to pursue a long time dream. Transferring chief financial responsibilities for the accounting department to John Reed, and, setting up his own financing, Mike McCaw built a golf course!

By 1987, McCaw's golf course, the Broken Tree Golf Course, Mason, Ohio was built and he returned to Belcan. His return was most timely: Belcan's explosive growth demanded McCaw's expertise in accounting and law and, importantly, he was available to help weather the financial crisis of 1988.

In February of 1988, McCaw moved into the executive offices on Anderson Way for a year. When business stabilized, he returned to Technical Services as Chairman. Also on this team at Technical Services were President Lane Donnelly, Vice President, John Kuprionis, and Cleve Campbell. The four contributed different skills to the separate corporation: McCaw, in accounting and law; Donnelly, in personnel management; Kuprionis, with twenty years experience in technical services; and Campbell, operating management experience in industrial firms.

Belcan Services Group is located approximately five miles from the Anderson Way home offices. In 1989, McCaw became Chief Executive Officer, Belcan Services Group, when it became independent of Engineering Services.

PROFILE OF JOHN ROY MESSICK

 When John Messick joined Belcan in 1984 he became one of Belcan's first process engineers at the time of its rapid growth. Later, he became Vice President of Process and Facilities, concentrating on the operations of the number of partnering relationships of the 1990s.

John Messick was born in Norfolk, Virginia in 1944. Upon the divorce of his parents when he was 13 he moved to Ft. Walton Beach, Florida where he attended high school. He graduated in December of 1968 with a Chemical Engineering degree from Auburn University, completing its cooperative program while working for St. Regis Paper Company in Florida. John was recognized as the outstanding engineering graduate and was bestowed with Phi Kappa Phi honors.

Because he was subject to the draft during the Vietnam War, he decided not to attend graduate school. He obtained a job in product development with Procter & Gamble in Cincinnati and worked there for four years. In 1972 he obtained a job with an engineering consulting firm in Cincinnati.

Messick worked up in this privately owned firm of about 20 people until October 22, 1984. He reached the position of second in command as Executive Vice President. After a disagreement with the owner, he was released and accepted the offer of Jim Young to help build up the process engineering area at Belcan.

During the period of the cash crunch in the summer of 1988, the new business with Lilly was a key savior. During this period, process engineering gained importance with the company.

Messick as Vice President headed the Process and Facilities division from the fall of 1988 to early 1992 when he was transferred to concentrate on the supervision of the growing number of alliances that Belcan had developed. Prior to the transfer, he was in charge of three engineering offices and more than five hundred employees. Gross sales were approximately $30 million per year.

John earned his masters degree from Xavier in 1992. As Vice President, he was a member of the Lead Team of Belcan Engineering.

PROFILE OF DANIEL CARROLL SWANSON

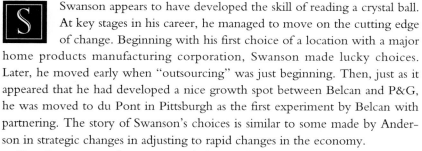

Swanson appears to have developed the skill of reading a crystal ball. At key stages in his career, he managed to move on the cutting edge of change. Beginning with his first choice of a location with a major home products manufacturing corporation, Swanson made lucky choices. Later, he moved early when "outsourcing" was just beginning. Then, just as it appeared that he had developed a nice growth spot between Belcan and P&G, he was moved to du Pont in Pittsburgh as the first experiment by Belcan with partnering. The story of Swanson's choices is similar to some made by Anderson in strategic changes in adjusting to rapid changes in the economy.

Dan Swanson was born on March 21, 1949, in Wise County, Virginia, literally a stone's throw from Kentucky. Unlike other engineers at Belcan, he did not need to use a cooperative program in his engineering education. Rather, he was fortunate to obtain a full scholarship for his four years in the one school that he had chosen as a pre-teenager, Virginia Polytechnic Institute (VPI). The son of a coal miner, Dan lived all but one year of his developing years in Virginia.

He entered college in 1967 and graduated in 1971. When he graduated, he did not worry about going to Vietnam since he had hit the jackpot in a most important lottery. His number was 334 out of 356. On September 4, 1971, after graduation, he married his high school sweetheart, Carolyn Maples, and accepted a job with P&G. He had a choice of locations: at the peanut plant in Lexington, Kentucky, or in the home office engineering division in Cincinnati. He accepted the one in Cincinnati. Since he was interested in the management track rather than technical, he went through P&G's "rounding and broadening" program. As a project manager for 10 years, with P&G in ten different jobs: designing, estimating, constructing, operating in two divisions, soap and paper. One year in the international division, he was in Japan (1974–75) to modify a plant for P&G international expansion.

Although his years with P&G were satisfying, in 1981, he accepted a position with Belcan. He could see that P&G with 3500 engineers was moving to outsourcing. By leaving the company early, he had better control over his choices. Having known Jim Young and Bob Smith from 1978 when Belcan was on Deerfield Road he knew that Belcan was the next place for him, therefore, he answered an advertisement for a project manager in early 1982. Dan fitted Young's plan of hiring a P&G person to deal with P&G and so was among the first to move from P&G to Belcan.

The move in 1982 was opportune for both Swanson and Belcan. Immediately, Swanson built up the business with P&G, especially in obtaining a new project for ALWAYS, a feminine protection product. Within a year, the project increased in size to several million dollars. Swanson feels that this project was the takeoff stage for Belcan in growth since it laid the foundation for Pampers project in 1983–85.

After ALWAYS was winding down, Swanson did not move to the bigger PAMPERS project. He was moved to a new client, Du Pont, in Pittsburgh to initiate Belcan's first partnering arrangement. Du Pont already had several partnering arrangements with other companies. Belcan's proposal beat out these others and started Belcan out in partnering. The new position had growing pains; Du Pont had had previous experience with partnering and well-defined company mores, but Belcan needed to adjust to the larger company. The result was Swanson's leaving after initiating the first successful partnering arrangement for Belcan.

Swanson returned to Cincinnati just when Jim Young was ready to strengthen the Industrial Products Division. Dan Swanson again fitted Belcan's needs and built the Industrial Products Division, and in 1993 became Vice President and a member of the Lead Team of Belcan Engineering.

PROFILE OF PATRICK WAGONFIELD

 Literally growing up with Belcan, Pat Wagonfield might be called the "all Belcan Boy," wearing many hats, using three different business cards, completing his college education after joining Belcan, and enjoying model building and video production.

Born in Darrtown, Ohio, on March 2, 1952, Pat moved with his parents and older brother to Hamilton where his father built houses. During his high school days, Pat worked for his father each summer. He began playing the guitar professionally when he was fifteen years old while at Hamilton's Taft High School. He attended the University of Cincinnati in its cooperative engineering program for three years while working as a contract engineer with Procter & Gamble.

In June of 1974, Pat joined Belcan while it was on Montgomery Road working as a contract engineer for Procter & Gamble in modeling and engineering. Soon Pat took over the model shop for Belcan and in March of 1980, he married Kathy Kley. His wife was a student at Miami University. The excitement of college and technology led Pat to return to college as well. This time he attended Miami and majored in systems analysis attending school at night and graduating in 1985. During this time Wagonfield operated Belcan's Model Shop on Kemper Road during the Pampers project for Procter & Gamble.

With his systems education and interest in computers, Wagonfield has focused on Product Design through modeling and then via CAD services. He built the model for the giesel, which was taken to Washington to secure development contracts with the U.S. government.

Pat enjoys tackling new challenges. His first interest is in industrial design, which leads to modeling and CAD. His most recent endeavor is the development of Belcan Industrial Video Productions, in which he designs the essential communication tools to move Belcan and its clients into the future. In short, when a project (particularly an *unusual* project) needs to get done, management knows that Pat can successfully and enthusiastically complete the project.

PROFILE OF JANE HOPKINS YORK

 After joining Belcan in 1984, Jane York became the voice of Belcan. To anyone calling the Belcan home office, Jane is more than a routine receptionist; she is the front line public relations promoter of Belcan. Jane's skills are extraordinary—polite, welcoming, and gracious. Callers to Anderson Way are charmed and provided a wonderful first impression of the company. Many Belcan employees have stated that their mornings begin on a positive note because of Jane York's sunny greeting. Less obvious is Jane's vigilance. Since several of Belcan's clients demand restrictions on information, a security system is essential. Jane is an unrelenting security guard, requiring Anderson Way visitors to sign in and wear conspicuous red badges. Throughout the day Jane keeps a watchful eye on lobby comings and goings. Jane knows the location of approximately five hundred Anderson Way employees and she has a phenomenal memory for repeat callers. Any deviations from the strict security system by executives, visitors, and even the owner are corrected immediately. Soon after joining Belcan in 1984, Jane York became what most succinctly could be called "the voice of Belcan."

Born on June 23, 1941, in Boston, Massachusetts, Jane York had a first career as a wife and mother of three children, Amy, born March 12, 1959, Brenda, born August 25, 1960, and Michael, born April 26, 1965. After a divorce on April 26, 1968, Jane began her next career as an office manager for dentists in Aurora, Indiana and in Cincinnati.

In October of 1984, Jane was employed by Belcan at the time of its rapid growth. After arriving at Belcan, Jane became a central figure as Manager of Support Services and as Communications Director. Her resume would include a number of activities such as hiring office staff, conducting training programs, Telephone System 75 Manager, involvement with an in-house magazine, *The Eagle Dispatch,* and serving as receptionist and supervisor of Belcan's phone directory. No matter how important these activities might be, the reason to include Jane in a separate profile is based on more than just rank and job descriptions.

When Jane considered moving from the Cincinnati area for personal reasons in 1988, she loyally remained at Belcan until the company's financial crisis had passed. When she left in the summer of 1989, many executives of Belcan not only wrote outstanding letters of recommendations but they called her in Florida urging her to return to Belcan. Six months later she agreed to return.

Jane York is an excellent example of the type of unusual employee cultivated by the friendly, informal climate created by an entrepreneur whose success is dependent on the people with whom he surrounds himself.

PROFILE OF JAMES H. YOUNG

 The gas turbine engine and Jim Young literally matured together; later, Young and rocket engines entered the space age together. With Belcan, in 1974, Young focused on designing problems of the giesel, but remained close to gas turbine engines with the Belcan GE Alliance in 1991.

Immediately upon graduating from Rose Polytechnic in June of 1951, he was hired by Allison and started to work on the J-33 gas turbine engine. On January 2, 1966 he became manager of Advanced Propulsion Systems at General Electric in Evendale, Ohio. With his expertise in engineering design he was a natural choice for Ralph Anderson to ask in 1974 to contribute to the design of the Belcan's revolutionary giesel.

Jim Young was born on a farm on August 2, 1923 near Rochdale, Indiana. His Hoosier family soon moved to Indianapolis where he lived for 37 years.

He graduated from Arsonal Technical High School in Indianapolis then volunteered for the Navy to become a pilot in World War II. However, physical exams resulted in his serving in the engine room of a destroyer-escort principally in the Pacific. He married in 1945 while in the navy; later he had two boys. He entered Rose Polytechnic in Terre Haute, Indiana in 1948 and completed the requirements in three years by June of 1951.

Jim was hired by GM's Allison Division to work on aircraft engines. The company had a large force of people who had built the piston aircraft engines during WWII, but it had just obtained the government contact to build the J-33 gas turbine engine designed by GE. In this first year, Young went to work on the J-33 becoming supervisor of the Analytical Stress Lab. In 1957 (after Sputnik) he became Manager of the Advanced Design Section, working on ramjets solar power, ICBM rocket motors, and other space age motors. In 1964 he was awarded a patent of an advanced light weight rotor system used in the space program. During this period, Young worked in the Cole Building with Anderson's Allstate engineers on stress tests of the T78. With Allison, Young remained in pure engineering but wished to obtain a broader perspective that was possible in a position with GE at Evendale.

Between 1966 and 1974, Young was employed by General Electric at Evendale, Ohio first as section manager, Advanced Propulsion Systems, and later as Program Manager, Advanced Engine Technology Department. In these positions, Young negotiated with the Air Force and worked on a variety of advanced systems, including vertical takeoff, lunar excursion models. In the early 1970s the funding for space and aircraft engines to General Electric was reduced. In 1974, Young was shifted to work on the MF 10,000 turbine engine for large industrial uses which he felt to be of less challenging efforts.

During a meeting with Ralph Anderson, he learned of the giesel and the problems that Anderson was experiencing with Bob Johnston. Jack Hope had moved to Washington. Young saw that it needed development work in its design. Young agreed to come to Belcan on May 1, 1974 with a deadline of July 1, 1974 set by the army. This deadline was met and continued development until 1977. After the grants for the giesel expired, Young became Belcan's Vice President of Engineering and oversaw the buildup of the in-house engineering until late 1975.

During the rapid expansion of the early 1980s, Young focused on the engineering projects. Young developed the Industrial Parts Department and in 1991 completed the contract for the General Electric Aircraft Engine Alliance which was located on the second floor of the Anderson Way Building. As the Senior Executive, Young remained Vice President throughout and was a member of Belcan Engineering's Lead Team.

FINANCIAL CRISIS AND CONSOLIDATION, 1986–1989

"Every setback in my life has ended up being for my better good."
RALPH G. ANDERSON

 At the beginning of the year 1986, Belcan had the characteristics of a basketball team that had just won the national championship! After its successful completion of the huge Pampers project for Procter & Gamble, Anderson could really enjoy the prospects for next year's game.

In 1959 Ralph Anderson was not quite sure where the next payment for his house would come from. Now, twenty-five years later, he had more than demonstrated the power of optimistic thinking.

Cincinnati newspapers reacted on the business page as they would about Pete Rose on the sports page: **"BELCAN PROFITS FROM RISK";** **"FAMILY-OWNED FIRM SOARS LIKE EAGLE"** were headlines in which the story described Belcan's purchase of Multicon and Lodge & Shipley quoting John Reed who had arranged the $10,000,000 purchase.

A photo from the *Cincinnati Post* in 1987 included a picture of three Belcan executives, W. J. Ashton, J. E. Campbell, and John A. Reed with the carved eagle in Belcan's lobby. In 1990, the eagle was the only one from the picture that remained at Belcan.

Other news articles in 1986 and 1987 concentrated on real estate developments with Chuck Kubicki who built Belcan's buildings and had been a close friend. By 1991 Kubicki, who did not respond when Anderson had cash flow problems, fell from close association with Anderson. Kubicki himself had some financial difficulty.

Reviewing the headlines five or six years later provides some insight about Anderson and the rapid changes that occurred in business. The headlines do reflect the facts of the time but they do not tell all the story.

Moreover, the "rest of the story" after 1989 is most interesting and demonstrates how a positive thinker can change the loss of several games to another winning season!

Furthermore, the personality of Anderson was graphically illustrated to the author in early 1988 as a guy who likes to win but does not believe that winning is the whole game; he can be relaxed and have fun when he is facing major problems. Yes, he likes to win but more basically, he just has fun playing the game!

Explosive Growth Leads to Disorganization

Anderson began to use management consultants to help him structure and staff the growing company. Chuck Comfort was retained for several years and helped locate several new key executives. At least two other management consultants successively funnelled advice to Anderson.

During the year of 1986, Belcan was attempting to fit Multicon into its structure. Gary Bates, who had joined the company in 1985, would leave it in August, 1986. William Thomas left the company in January of 1986. Michael McCaw, who had been made Chief Financial Officer in 1985 would leave the company in June of 1986 to build a golf course in the Cincinnati area. The renovation of the mansion at Walnut Hall on the farm in Mercer County was underway.

A company in the rapid stage of growth tends to outstrip the means for coordination. Also, a small company that depends to a great extent on a single client can lose control of its own company. Anderson had proven in the past his uncanny ability to assess the qualifications of new recruits. Yet, in 1986 he relied on the expert advice of management consultants. In short, he attempted to shift from past successful approaches of entrepreneurship to a generally accepted "school solution" management approach that did not fit his strengths.

The realistic picture of 1986 is not a pleasant one. There were changes and departures. It was confusing for all participants at the time and a restatement of the past in an orderly framework would be misleading and unrealistic. No organization chart (even a monthly one) could show the real flow of authority. Ralph Anderson was chief executive officer and all executives had access to him. At this point in Belcan's history, things were *unorganized,* but not necessarily *disorganized.*

The organization of Belcan was stated in a brief memo in August 1986. Anderson was CEO with four executives reporting directly to him: John Reed, Controller, Jim Young and Lane Donnelly in Technical Services, and Cleve Campbell in Engineering, Field Offices, Multicon, Construction, Sales, and Advanced Technology. It became evident to Anderson that there were too many operations to be handled by one man and thus, there was a need for a regrouping and reorganization.

Executive Recruitment (1986)

On February 3, 1986, Cleve Campbell was recruited to handle Human Resources. (See *Profile: Cleve Campbell*.) Campbell was an engineer, but had experience in directing human resources and line management in mining and manufacturing in West Virginia. His lack of specialization would make him most valuable when a general purpose fireman might be needed.

On May 5, 1986, William Ashton from the auto industry in Detroit became Vice President of Sales and Marketing. He had experience in marketing campaigns of large companies. The availability of Ashton had been made known to Anderson by Charles Comfort, management consultant. William Thomas who had been recruited to expand sales had left the company in January of 1986.

On June 2, 1986, Dr. J. J. Suarez came into the company as Vice President, Engineering. (See *Profile: J. J. Suarez*.) Suarez had both academic and industrial experience in engineering. After waiting six months, J. J. recruited Barb Schmidt as secretary (See *Profile: Barbara Schmidt*). Also in June, Henry Helm was attracted to Belcan by several old friends already with the company. (See *Profile: Henry Helm*.)

On June 16, 1986 John Reed was brought into the company from the public accounting firm, Touche Ross, to be controller of the company. Reed already was familiar with Belcan's finances over the previous six years since he handled Belcan's account as a public accountant. Reed was the first certified public accountant to be with Belcan. His experience had been in public accounting with less experience in company or managerial accounting.

On November 10, 1986, Ronald A. Beymer was made Vice President, Development. (See *Profile: Ron Beymer*.) Anderson used the term development instead of marketing because he had had poor experience with "marketing" people. With his strong feelings toward satisfying the customer, development was his term when thinking of "partnering relationships or alliances."

Management attention was stretched to the extreme. For a time between November 1986 and February 1987 Anderson assumed operating supervision of engineering himself when Campbell was assigned direction of Multicon, Manuflex, and other parts of the business. New managers were being brought in so rapidly that it was difficult to maintain the smoothly running operations demonstrated during the earlier years of Belcan's operations.

Formation of BGP Services, a Partnership

The year 1986 also saw the conclusion of the Pampers project with Procter & Gamble and the extended negotiations to form a partnership with GPA and Foxx Associates (PDX) to continue design engineering for the Paper Division of P&G. In November, BGP began operations with 304 employees, with approximately 200 from Belcan, seventy from GPA, and thirty from PDX as a separate partnership with Vincent Salerno (See *Profile: Vince Salerno*) as Chief Executive and General Manager. The loss of the 200 Belcan engineers to BGP unfortunately occurred just as the Anderson Way major expansion was being completed. The birth of BGP was particularly painful for its parent Belcan. In Chapter Seven we examine the successful development of BGP over the next five years when we consider the other Belcan Alliances, but we now observe its unfavorable effects on Belcan and the powerful role that Procter & Gamble played in its formation and operations.

Cleve Campbell had been recruited through Glen Baldwin Management Resources to be a human resources manager. Immediately upon arriving in February, 1986, he became the fireman to put out the fire at Belcan. He first was used to handle a problem in Pittsburgh that involved the relations between Du Pont executives and Belcan executives. Problems had developed in the relations between Belcan and its client in a novel long term relationship, known as *partnering,* or *alliances.* Campbell had few models for such activities; the concept was just being developed and he found himself facing the two doors of the lady or the tiger syndrome. He was either on the leading edge of excellent opportunities or on the brink of disaster.

Then, on August 27, 1986, Campbell was appointed General Manager of the Engineering Division to replace Gary Bates who had been moved to the new partnership, BGP. Campbell uncovered a number of problems which needed attention. In the effort of expanding and keeping up the momentum from the Pampers project, projects had been obtained with which the company had little expertise. In the words of Mike McCaw, the company "took a bath" with regard to some of these projects.

Belcan Becomes a Holding Company with Subsidiaries (1985–1987)

Things looked dismal. The morale of the company had hit the lowest point in its history. Small matters such as the parking availability on Anderson Way, became inflated problems.

Because the growth of Belcan created new liabilities for the owner, lawyers and accountants recommended that the activities be contained within separate corporations. During this time, Belcan became a holding company with a number of operating companies, a move that answered some of the legal and financial problems of the company, but made the human organization and management more difficult. Without listing the names of the new corporations, we can summarize, between April 20, 1985, and February 13, 1987, twelve new corporations were formed. Up until this date, only one corporation had been formed—the original Belcan Corporation formed in 1958.

It appeared that Belcan, an engineering company with 90 percent of its professional personnel in engineering, had been overtaken by legal, financial and management experts who handily lost sight of Belcan's primary business: providing technology and engineering services to its clients. As a result of this mismanagement or perhaps, overmanagement, the foundations of favorable cash flow were greatly disturbed.

Lodge and Shipley Tool Company

In January of 1987, Belcan acquired one of the oldest machine tool companies in Cincinnati for $10,000,000 and placed its assets in Manuflex, a corporation that had been formed on December 12, 1986. The company had a large amount of land and empty buildings in an older portion of town located nearby downtown Cincinnati. From the first, its location presented a problem: it was ten miles from Anderson Way's Blue Ash location.

The reason for the purchase seems logical. Anderson learned that Lodge and Shipley had been trying to sell General Motors a turning cell for its gas turbine operations. Since Belcan had had close association with General Motors, Anderson felt that Belcan could pursue making these turning cells for GM if Lodge and Shipley were purchased by Belcan.

Following the purchase, Belcan was successful in producing sixteen turning cells for General Electric in Wilmington, N.C. However, manufacturing of products required different management skills from servicing with engineering expertise. Furthermore, producing products could not be financed in the same way that providing engineering services could. Lodge and Shipley had orders on the books from Ford for turning cells and from others for up to $40,000,000, but its bank (Star Bank, formerly First National Bank), would not lend the money necessary to finance the production. In the auto business, 90 percent of the payment from the purchaser is not received until the product is delivered. Thus, a firm needs to secure bank financing for working capital.

Belcan had been using Ameritrust as its bank and Ameritrust would not become involved with the financing of Lodge and Shipley. Anderson found himself in a tight squeeze between the banks and moved to sell parts of the business in order to increase liquidity. The end result was that money paid for Lodge and Shipley was lost and Belcan was forced to reorganize. Again. Cleve Campbell was instructed to close out Lodge and Shipley.

Campbell came up with a number of potential purchasers of Lodge and Shipley only to have each of them fail to be completed due to lack of financing. Finally, he was able to sell the patents and inventories to a group of venture capitalists and the physical assets to a Chicago firm who specialized in moving buildings.

In September 1987, through Belcan Technical Services Division, a group of engineers were hired from Acme Company in Cleveland, Ohio. Belcan's opportunity for these engineers resulted from Acme's union and financial troubles. One of these, William Waddell, was named General Manager of Manuflex, Inc. to handle the manufacturing operations housed in Lodge and Shipley's buildings. The others remained in Cleveland.

In 1987–88 Waddell directed two manufacturing operations of Manuflex in these old buildings: Waterjet cutting system of leather and Ulticon, Inc. Waterjet cutting system of leather was a technological advancement for the shoe industry and it had several advantages. It is ten times faster than other systems of cutting leather, thus saving on raw material costs. The chief problem was to start up sales for a system that did not fit the distribution channels formerly used by Belcan.

A new company, Ulticon, Inc. was organized to produce a patented product, 50 percent owned by Anderson and 50 percent by Jerry Yankoff. The subsidiary produced a revolutionary method of cutting metals involving liquid carbon dioxide that resulted in a thermal method of cutting metal. The usual troublesome spiral of metal shavings was eliminated by breaking it down into small bits.

Ulticon was born at just the time that management attention was spread thinly and funds were not available for development. Nevertheless, it had several advantages that at some future time could result in profits: the finish of the metal resulted in a polished product. The handling of trash was reduced, lowering labor costs. It was thought that with additional development, this patent could become valuable.

Anderson retained the title to the Ulticon patent and kept it on the shelf for future use when funds became available. However, other opportunities for the use of the funds continued to appear to be more promising. Further exploitation of the patent was delayed.

Cash Crunch: Summer of 1988

The rapid growth of the early eighties, the diversification in the mid-eighties, and the shift to manufacturing products in addition to providing clients with expert engineering set the stage for a cash and credit crunch with Belcan's banks in 1988. The underlying causes of the crisis began with several adverse developments beginning toward the end of 1986.

First, the profitable Pampers machinery design job for Procter & Gamble provided funds to Ralph Anderson for expansion into a number of enterprises rapidly. The new activities outgrew the organization structure to handle the management functions; in other words, the company bit off more than it could chew.

Second, after its start in 1976, full service engineering expanded so rapidly that it was hard to build space to keep up with the needs. The new building on Anderson Way became too small, therefore, it was doubled in size, with the space becoming available just at the time the need was lessened. By 1987, the expansion in space did catch up. The headquarters building had space availability that remained vacant for a period of time. Resulting was an increase in cost of operations and an operating loss.

Third, the separation of Pampers into BGP operating independently of the parent company took 200 people out of the headquarters building. The client, Procter & Gamble, made it clear that the separation was prerequisite for continued work for the paper product division.

Fourth, the failure to obtain sufficient financing to continue operations of the Lodge and Shipley property required that the assets be sold. This sale resulted in a large loss (between $5 and $10 million) that had to be swallowed immediately.

Fifth, the experiment with the Ulticon process showed promise, but its development required large amounts of funds for development. Belcan did not have such funds at this time.

Sixth, the delegation of cash planning and working capital management to the controller, John Reed, caused Anderson to be surprised by the seriousness of the cash flow problems. The banks involved (Ameritrust and Star) had lost direct contact with Anderson and had serious questions about dependability of John Reed's statements.

Seventh, conflicts arose between sales and project managers. William Ashton had increased sales significantly; however, sales were being made without proper regard to the available skills to handle the projects. Project managers were faced with expenses exceeding revenue. Without close cooperation and coordination between sales and project managers, Belcan profits were hurt even though revenue continued to increase. After the summer cash crunch, reorganization was needed. On October 14, 1988, William Ashton left the company; his position was eliminated with the reorganization of late 1988.

Finally, so many problems hit at the same time that the need was to return to "sticking to one's knitting"—for Belcan that knitting was to provide clients with engineering services, both temporary and full service.

Retrenchment and Cash First Aid

The financial crisis continued over a period of several months. Belcan's bank, Ameritrust, set strict limits to the actions of company management. No funds were to be used relative to Lodge and Shipley. Star Bank would not approve more loans to Belcan since it had been the bank for Lodge and Shipley.

Belcan's payment of bills was stretched as far as possible. During the period of searching for a buyer for Lodge and Shipley, the financial situation looked grave. At one time it was suggested that Lodge and Shipley be taken into voluntary bankruptcy; however, even if it might have been a way to save funds, Anderson would have no part of any attempt to use this route. He had always valued his integrity and he considered this route to be a compromise of his integrity.

However, alternative routes were perilous. After seeking loans from a former close real estate partner and being turned down, Anderson had to contact an old friend in Harrodsburg. Through the help of a small bank owned by this friend, Anderson was able to raise $60,000 over a weekend to keep his head above water. Other short term measures were adopted to work from week to week but the only lasting solution was to increase profits from Belcan by returning to what it had always done well in the past.

Reorganization

In October of 1988, the corporate structure was simplified. Belcan Services Group was separated, and it was housed in the complex built by Anderson on Kemper Road.

Belcan Engineering, Inc. was formed to concentrate on in-house engineering and to develop the new concept of partnering and alliances. J. J. Suarez became president with his office located at the headquarters address, 10200 Anderson Way.

Ralph Anderson and his new controller, Paul Riordan, (See *Profile: Paul Riordan*) formed the holding company for the two operating companies. John Reed, the previous controller, left on January 13, 1989.

At the beginning of this chapter, we saw that Belcan was in a boom stage of its cycle. During this cycle, acquisitions provided the means for diversification. In order to finance this growth through acquisitions, it was required that Anderson continually give attention to commercial banks for their support. Commercial banks during this period called the shots for Belcan and we see that Anderson changed banks more rapidly in the 1980s than he did when the company was smaller (See Table 6–2).

Table 6–1

ACQUISITIONS BY BELCAN (1978–1993)	
1978	Wilde and Krouse
1985	Multicon, Inc. (Founded by Flisik and J. E. Campbell in 1982)
1986	Manuflex Corporation was incorporated by Belcan as successor to Lodge and Shipley. Ten Engineers, formerly of Acme, Cleveland, became employees of Manuflex in production with machine tools patented by Anderson
1987	Ulticon (Jerry Yankoff)
1992	McGraw Engineering, Middletown, Ohio, Custodio, Suarez, & Associates (CSA)

Table 6–2

COMMERCIAL BANKS USED BY BELCAN	
1963	Provident Bank
1969	Star Bank (Formerly First National Bank)
1971	Fifth Third Bank
1982	Winters Bank of Dayton, Ohio (bought by Bank One, of Columbus, Ohio)
1986	Ameritrust (Cleveland) became Belcan's bank
	Star Bank (First National of Cincinnati had been Lodge & Shipley's Bank)
1991	Central Trust (PNC)

Note: The research on Belcan, as we described in the Preface, depended on interviews with personnel from Belcan and associates of Anderson on the outside concerning detail of the company activities through 1987. At that time, the author obtained direct observations of the company through several months of consulting with Belcan, soon after retiring as Alumni Professor of Management, College of Business and Economics, University of Kentucky.

In 1987, the author had been introduced to Anderson by Professor David Blythe who had taught Anderson in the College of Engineering. The initial observation of these visits to Blue Ash was to make oral history interviews of Anderson for the archives of the University of Kentucky.

The first tape, made on March 30, 1988, is on file at the UK Library. This interview covered Anderson's life to approximately 1980. During the first three months of 1988, the author interviewed most executives of Belcan in a consultants capacity and discussed issues with Anderson.

A major reason for mentioning these facts in this chapter is that by hindsight, the author is amazed that Anderson could have been so much at ease and open to interviews during a period in which he had big problems digesting the activities during the early months of 1988. Anderson really was having *fun* meeting challenges and was confident that things were going to turn out all right even though at the time most people would have shown considerable stress.

PROFILE OF RON BEYMER

R Ron Beymer has unique characteristics among Belcan's officers: he is from Iowa, he is an industrial, not a mechanical, engineer; and he was not a part of a cooperative program, although he paid for most of his education himself. He preferred marketing and sales over the technical.

Beymer's father operated a bulldozer briefly in Orange County, California. Ron was born there on March 20, 1949. The family moved back "home" to Diagonal, Iowa, where the family operated a propane and service station business.

Attending local schools in this small midwestern town, Ron was active in school sports. The school is still celebrating its 1938 basketball team championship. Before graduating from high school in 1967, he spent a summer at the University of Iowa at Iowa City in pre-engineering. He chose Iowa State for his college career since he felt its industrial engineering program was superior. Although he did not participate in the cooperative program at Iowa State, he paid for 75 percent of the four years college expenses.

Immediately upon graduating from college, Ron married Kathleen Jo Sokotka on July 17, 1971 and accepted a marketing position with Johns Manville. After the company moved him to Chicago, he completed his MBA at Illinois Institute of Technology in 1975 and became area marketing manager in asbestos concrete pipe for water treatment plants.

In 1982, Beymer transferred to Defiance, Ohio, where he was production supervisor when Manville went into Chapter 11 bankruptcy. After the bankruptcy, he became plant manager of a carton plant.

With the uncertainty from the Chapter 11 operations in 1986, Beymer joined Belcan. Since Ron wanted to get back into sales, he worked closely with Bill Ashton, Belcan's Sales Vice President at the time. During recent years, Beymer has concentrated in sales and moved into a key position of building a new partnering arrangement central to Belcan's 1990 strategy.

PROFILE OF CLEVE CAMPBELL

 Becoming a family man while in college, Cleve's entire career has been determined by the better interest of his wife and children. He worked several jobs while in college, moving from a management track at a large industrial company because his son developed cancer and his wife was seriously ill during the childbirth of their third child. He chose to stay in Cincinnati to work for Belcan because of his daughter's preference to stay—all these family factors determined his employment choices.

Campbell's entire life gave him wide experience in industry as he worked in mines while living in coal camps, on a management track for an integrated steel company, and in management of a number of companies in a consulting firm. His experience before joining Belcan was most varied and general.

Campbell was born on December 17, 1942 in a coal camp in Price Hill, near Beckley, West Virginia. Campbell's mother's family came from Scotland in the 1600s. Cleve's father, well known in the electrical mechanization of mines, worked for several large mining companies in West Virginia, Utah, and Pennsylvania. Cleve did not live outside a coal camp until he went to college.

He graduated from Pennsylvania State University with an undergraduate degree in Economics and Engineering in 1965. His first job was with Jones & Laughlin Steel, Cleveland, Ohio, on the special management training track in which he received experience in all aspects of the company and worked his way up to fourth level manager (youngest at age 24) before leaving in 1972 due to the illnesses of his wife and son.

From 1972 to 1984 Campbell was manager of Industrial Relations and Safety for Pickand & Mather & Co., a management company that managed eight corporations in Ohio, Kentucky, West Virginia, and Australia. During this time he, along with three other managers, directed expansion from 150 to 3500 employees.

In 1984 Campbell moved to Cincinnati to become Vice President of Human Relations for a new corporation formed by the Italian company, ENI and Occidental Petroleum, named ENOXY. After a change in the CEO of the new company in 1987, it was decided to transfer the headquarters to Canton. Campbell was one of two employees out of the 80 in the Cincinnati office who were offered better positions in the larger company but the requirement was that Campbell would have to move from Cincinnati.

Because Campbell's daughter was sixteen years of age, a difficult age to move, he declined the offer but was assigned the job of terminating all the others and closing the company. He then registered with Keith Baldwin, an executive placement company, and on Feb. 3, 1986, Campbell went to Belcan as Vice President of Human Resources. Campbell held various positions at Belcan until 1992 when he was appointed President of Belcan Temporary Services.

It is interesting to note that Campbell was hired in both ENOXY and Belcan as Vice President of Human Resources but in neither company did he serve in that capacity; however, in both companies he served to operate and terminate companies. After serving as fireman for various problems in Belcan, he was assigned to find buyers for Lodge and Shipley in 1989.

PROFILE OF CHARLES EDGAR EASLEY

 Chuck Easley is a chemical engineer who has concentrated on the technical aspects of process design. This background has helped him understand customer needs as leader of Belcan's Chemical and Process Program.

Chuck Easley was born on June 16, 1939 in Ft. Wayne, Indiana. He was raised on a 60-acre Hoosier farm. Chuck went to rural elementary and secondary schools (grades one and two were in a one-room schoolhouse where he first met his wife, Reta). He majored in chemical engineering at Purdue University on a full scholarship through the Naval ROTC and served five years on active duty with the Navy in Admiral Rickover's nuclear propulsion program. Upon release from the Navy, he accepted a position in development at Procter & Gamble where he worked on Pampers to change from creped paper to the lower cost fluffed paper pulp which is used today.

After four years at P&G, Chuck wished to increase his scope of experiences and joined Katzen Associates, an engineering firm in Cincinnati. He handled various projects dealing with ethanol and the scrubbing of gases to recover sulfur dioxide for use in wood pulping. While at Katzen Associates, Chuck worked with John Messick who later joined Belcan. Next came several years with Pedco, another Cincinnati engineering firm, where he worked on processes for producing ethanol for use as a gasoline additive. In 1985 he returned to Katzen for another five years before joining Belcan in February of 1989.

In an early assignment with Belcan he led a successful crash team effort to calculate expected air emissions in support of permit applications for a client's major solvent recovery project. Later, he became process and controls manager and then director of the process and chemical program with a focus of sales and management.

Chuck is married to Reta Gerig, his childhood sweetheart. He has two children from a former marriage. His daughter graduated from Miami University and his son graduated from Cornell. Both reside in Cincinnati.

PROFILE OF ARTHUR E. FROHWERK

As a client of Belcan representing Disney in California, Frohwerk became impressed by Belcan's quality of work and joined Belcan in 1989. Born on the west coast of the United States, Art was unfamiliar with the Cincinnati area until he joined Procter & Gamble when he completed his engineering degree at Harvey Mudd College in California.

Born on September 25, 1950, in Portland Oregon, Art showed an early technical interest, making a flashlight from a band aid box when he was five, and in high school, winning an award by assembling a TV camera, and advanced in boy scouting to Eagle rank with 31 merit badges.

Frohwerk obtained a scholarship in engineering at Harvey Mudd College, in Southern California. Art was attracted to Harvey Mudd by its rigorous program, a General Systems major with opportunities in humanities and psychology, and its small class sizes permitting teachers to serve as mentors and individual coaches. While in college, Art met Jo Penny, a student at Pamona, whom he married on August 28, 1973.

After graduation in May of 1972, Art was recruited by Procter & Gamble as project manager and as gatekeeper who introduced new technology from one division to different applications in other divisions. His division manager encouraged him to have a big brother, an appealing idea to Art with experience with Harvey Mudd mentors and Boy Scouts. The gatekeeper role required him to seek new technology, often found in Southern California, enabling him to see his fiance more often. While at P&G, Art served in diverse capacities. One year he was project manager of Pringles in Jackson, Tennessee. His gatekeeper function served varied products, such as Pampers, Rely, soap and food.

Two daughters were born in Cincinnati, Jennifer in 1977 and Alison in 1980. The completion of his family added new dimensions to his aspirations affecting his job. After seven satisfying years at Procter & Gamble, the Frohwerks sought job opportunities on the west coast so that their two children could grow up near their grandparents.

Art joined a privately held Disney company which later became Walt Disney Imagineering. As chief engineer and department manager of Show/Ride Systems, he directed the installation of equipment for Epcot in Florida and Disney in Tokyo. Later, as a department manager with 215 engineers reporting to him, Art reduced the Disney engineers to 26 by outsourcing with Belcan. Belcan opened a Glendale office. John Kuprionis and Ron Beymer worked closely with Disney.

After a top management shakeup at Disney, Art found himself without a job. He considered the role of entrepreneur, but in October of 1989, joined Belcan as project director in Process and Facilities. He focused on low-margin, high-volume consumer products. In 1990 he helped develop a partnering arrangement with Eli Lilly called Boomer. It involved the translocation of five existing international capsules—making operations into two, one in Spain, the other in the United States.

Profile of Henry Helm

Henry Helm is the personification of Belcan's diverse managers. He is of German ancestry born in the area now known as Romania. His wife and children exert direct impact on his professional choices. He is a born-again Christian and has faced extreme risk in having children.

Helm was born on March 3, 1935 to German parents who had been among one million Germans moved by the Austro-Hungarian Emperor to be farmers in the fertile fields along the Danube. In 1944, when he was nine years old, his entire family moved west with a 24-hour notice to Budweis, Czechoslovakia and then to Braunau, Austria, to avoid occupation by the Russian Army. After graduating from high school in Austria, he was sponsored by an aunt to immigrate to Cincinnati in November, 1951 (he was naturalized in March of 1956). He secured employment with a company which made aircraft instruments. During evenings he took classes in mechanical engineering at the University of Cincinnati.

Henry was drafted and assigned to a combat engineering company near Heidelberg then transferred to 7th army headquarters near Mannheim. At this time he met his wife, Eleonore Koehnen, married three months later on December 19, 1959, and returned to the United States in May of 1960. He rejoined C&I Girdler which sent him back to Germany as a project engineer.

His first son, Andrew, was born on September 19, 1960 and Daniel was born on September 24, 1961. Eleonore, a nurse, knew that with her Rh problem there was great risk in having a third child. She and Henry agreed to take the chance. After returning to the U.S. in 1964, a third child, Jennifer, was born on May 14, 1966 in perfect health.

In 1966 Henry went to Calgary, Canada for a large grass roots fertilizer complex in an early partnering venture with the Canadian Oil Company and Farmers Co-Op. The company reorganized; engineering went to Louisville with corporate management, sales and commercial development consolidated in the Cincinnati headquarters. Given a choice between the two, and in spite of his engineering qualifications, Henry chose Cincinnati because Eleonore wanted to live in the German atmosphere of Cincinnati. Bechtel bought Girdler in 1970 and closed the Cincinnati office.

Still wishing to remain in Cincinnati, Helm took a position with AM Kinney until 1982. He moved to Wilson, North Carolina to join an ethanol venture. In June of 1986 he joined Belcan. Both Ron Rountree and Charlie Reed of Belcan had worked with Henry previously at C&I Girdler in 1970 and thus were important to his move to Belcan.

At Belcan, in 1989, Helm was selected by Larry Loll of Du Pont for the Belcan/Du Pont Regional Engineering alliance in Pittsburgh. Soon Loll was transferred to Wilmington by Du Pont and another manager was transferred to be Helm's counterpart at Du Pont. As a result, in May of 1991, Helm returned to Cincinnati to become program director for pharmaceuticals. With his previous contacts with Bob Hipple of Eli Lilly, Henry helped Belcan develop the partnering alliance with Lilly.

PROFILE OF
WILLIAM STEWART McKNIGHT

William Stewart McKnight is an unusual example of a manager who came to Belcan; he is an engineering manager without an engineering degree; he was an early specialist in operations research and management systems, receiving his education from a liberal arts college. He grew up with the steel industry in northeast Ohio and developed a reputation in non-destructive testing and managed several teams with General Electric who obtained patents on testing sensors for factory automating processes. Before arriving at Belcan just before its cash crunch, he had already survived the organizational deaths of several companies and a sick industry.

McKnight was the eldest of four children, born in the highly industrialized valley of Youngstown (Canfield) Ohio, on October 31, 1941. Bill's mother died when he was eleven. Bill's father was formerly a farmer, later to become a blast furnace foreman.

Working on farms each summer, excelling in sports (pole vaulting and football), McKnight also had a scholastic record in mathematics and the physical sciences that enabled him financially to graduate from Westminister College, New Willington, Pennsylvania, a small private liberal arts college in 1963. Although Westminister had a joint program with Carnegie Mellon in engineering, the financial pressures were too great to pursue engineering.

In 1963, McKnight joined Youngstown Sheet and Tube in its inventory process control department. Using early computers, he became known in the field for non-destructive testing and quality assurance in a research facility that was shut down in 1971. Bill worked for one year at the Hoover Corporation before he was asked to return to the steel company to form and head a team for preventive maintenance and acceptance testing for the company. In 1977 the demise of Youngstown Sheet and Tube and its takeover by Lykes Steamship Company resulted in many layoffs, but Bill continued to advance, despite this uncertain environment. With his marriage to Marilyn Carano on June 24, 1967 and the birth of their daughter, Kelly, in November 1974, Bill was ready for a more stable environment. Through his reputation in non-destructive testing and using the help of a head hunter, he was offered a job with General Electric.

On January 17, 1978 Bill became manager of non-destructive testing of field run engine components at General Electric at Evendale, Ohio. His son was born on August 23, 1978. Bill obtained his MBA in May of 1984 through the executive program at Xavier University in Cincinnati.

While at GE, McKnight built and managed several teams of high tech engineers who obtained several patents on testing sensors for the factory automating process. His work with Dave Godfrey, who moved to Cleveland with Acme and then to Belcan during Belcan's rapid growth in 1986, introduced him to Belcan.

On March 17, 1988, during the "shakedown" of Lodge and Shipley, Bill came to Belcan from GE as Jim Young began to implement the alliance with GE. Young supported Bill's interest in building a unit in Belcan to sell computer and electronic systems to clients as a part of the Industrial Products Division. It was this branch of Belcan that McKnight expanded into an important activity for Belcan.

PROFILE OF PAUL RIORDAN

 Arriving at Belcan on January 9, 1989, Paul Riordan, with extensive experience in both public and private accounting, became chief financial officer after the cash crunch of 1988. With his all-business attitude, he advises Anderson on accounting, tax, law, and financial matters. As the Chief Financial Officer, Riordan is the only executive, other than Anderson, in Belcan Corporation (the holding company) in this capacity. He is responsible for the accounting in the subsidiaries and maintains the books for Anderson Circle Farm in Kentucky.

Paul was born on January 19, 1944 in Dayton, Ohio, to a mother from Louisville, Kentucky and a father from North Vernon, Indiana. His father, a graduate of Purdue was an engineer with Frigidaire. Later, the family moved to the Cincinnati area where his father was an engineer with Crosley. When he was eighteen, his father died, while he was in his first year at the University of Cincinnati. He and an older brother helped their mother educate a younger brother and sister by working in part time jobs while continuing to major in accounting. During the last year of college in 1966, he secured a job with an accounting firm which gave him increasingly complex tax problems.

Riordan was married on June 12, 1967 and continued his education with a Masters Degree in Management at Xavier in 1968. Between 1967 and 1973 with Ernst and Young, he received opportunities to handle such accounts as Cincinnati Milling Machine with its 5,000 employees. In 1973 he moved to Andrew, Littner, Mancini & Vogel, a large local CPA firm.

In 1977 Riordan became vice president of finance of Miami Margarine Company where he worked with President Bob Clark (See *Profile: Bob Clark*). In 1986, Riordan completed the M.S. in Taxation at the University of Cincinnati.

The Riordan family lives in Indian Hill where he is the clerk/comptroller of the Village of Indian Hill.

PROFILE OF VINCENT SALERNO

 In November of 1986, Vincent Salerno was chosen by the three partners of BGP and its client, Procter & Gamble, to be General Manager and Chief Executive Officer of BGP Services, Inc. The selection of Salerno and the organization of a separate company was determined by BGP's customer, Procter & Gamble.

Salerno grew up in New York City, completing the Bachelor of Mechanical Engineering in 1957 from CCNY. He completed his Masters Degree in 1963.

From 1957 to1981 he worked for the Scientific Design Company, Inc. in New York moving from Project Engineer to Senior Vice President and Chief Operating Officer in 1977. In 1978 he moved to a related company in Houston, Texas, as President.

In 1981 he moved to W. W. Kellogg Company, Houston. In 1982 he was CEO of Bentach Engineers, Inc., Pasadena, Texas. From 1983 to 1986 he was Vice President of Petroleum and Chemical Engineering Division of Brown & Root, Inc., Houston.

Throughout the 30 years of practical experience, Salerno completed his MBA from Pepperdine University, and took short courses for training in Total Quality Management (Deming), Quality Improvement Program (Crosby), and Making Quality Happen (Juran).

After Ralph Anderson bought out the other two partners of BGP in 1990, Salerno was appointed Vice President, Belcan Engineering Services Group and reported to J. J. Suarez.

PROFILE OF BARBARA SCHMIDT

Barb Schmidt is an enthusiast of positive thinking. Anderson gave her the book, *THE MAGIC OF BELIEVING,* by Claude M. Bristol, in 1992. She contends that her success in country and western dancing competition is due primarily to her mental reaction to the book. As of the end of May, 1993, she and her partner, Richard Metzger, since June of 1992 had competed in Dayton, Ohio; Knoxville, Tennessee; Harrisburg, Pennsylvania; Atlanta, Georgia; Louisville, Kentucky; South Bend, Indiana; and Chicago, Illinois. In each of these competitions they placed very high, and in the Louisville and South Bend Competitions they won the championship for their Division. Barb's mentor was as proud as she was!

Born in Fort Wayne, Indiana, Barb moved to Cincinnati in 1963. She has one daughter and two grandsons. In 1987, Barb answered a Belcan newspaper ad and secured an interview with Dr. J. J. Suarez on January 28, 1987.

While Barb was waiting for her appointment in the lobby of Belcan, she noticed a "well-dressed, distinguished man" come into the lobby, proceed to the stairs behind the statue of the eagle, and dust the carving with a feather duster. Not until later did she learn that the "cleaning man" was the owner of Belcan!

In the interview, she was asked by Suarez what the first thing would be that she would do for him if he hired her. Indicating a stack of papers piled high on his desk, she replied that reducing the stack would be the first challenge she would undertake. Schmidt and Suarez were immediately compatible, so she started the next day, January 29, 1987.

Humor among the executives immediately appeared. For example, on one occasion, while Barb was placing a large number of papers into a notebook for J. J., she commented that they all would not fit. J. J. insisted that she could if she "squizzed" them in. Barb replied, "O.K., I'll 'squiz' them!" Overhearing this and other such incidents, Cleve Campbell began to refer to Barb and J. J. as "Lucy and Ricky", comparing them to Lucille Ball/Desi Arnez characters. Having worked with other members of Belcan's Lead Team, she noticed that Anderson's sense of humor was contagious. The humor appeared to cement the team effort at the top and then flow to other levels of the company.

Barb arrived at Belcan as it was weathering the difficulties in 1987–89. When Suarez moved from Vice President of Engineering to President of Belcan Engineering Company, Barb moved with him, helping to implement the new strategy. Barb continues to serve as J. J.'s best sparring partner and invaluable assistant.

Profile of Benjamin Spidalieri

 Born in Italy during World War II, Ben Spidalieri became a naturalized U.S. citizen in 1962. Ever since, he has developed a lifetime interest in the lamp division of General Electric, first as a direct employee of GE and then in 1987 as a permanent employee of Belcan with the title of Vice President of S.E.E.D. and its work with the Lamp Division.

Born in Campobasso, Italy, on November 30, 1944, Spidalieri went through the fifth grade in Italy speaking only Italian. In 1955, Ben's father, a cabinetmaker, his mother, two brothers, and two sisters immigrated to Pittsburgh, Pennsylvania where Ben's uncle was in contracting. Because no one in the family could get a job in the Pittsburgh area, the entire family moved in 1956 to Cleveland where Ben attended a Catholic school. He picked up English from talking with his friends.

Although he was offered a part fellowship to Case Institute, he entered the cooperative program at Cleveland State University in electrical engineering in 1963 because of financial reasons. In 1964, as part of the cooperative program, Ben joined General Electric in their drafting department and later worked in two smaller firms in Cleveland before obtaining his BS in electrical engineering in 1968.

After graduation from Cleveland State, Ben joined General Electric in their Lamp Division. As a part of the Edison Engineering Program, Ben rotated in a number of departments including product design and liaison with manufacturing and the introduction of new equipment. Later he was promoted to Manager of Equipment Concepts and Design for the Lamp Division.

Spidalieri was able to take advantage of the strong education orientation of General Electric by attending courses in management, manufacturing engineering, and Quality Course in the GE Crotonville Institute, a highly respected top management course.

In the early 1980s General Electric began to think in terms of making agreements with other firms to perform many staff and functional activities including engineering. In the spring of 1987, GE management contacted Belcan's Anderson and Suarez about a partnering arrangement.

GE had identified Spidalieri as a GE manager who could head up an alliance between Belcan and GE Lamp Division. During a 10-day period, Spidalieri made decisions as to which engineers would be moved to Belcan and was liaison for the agreement between the two companies. At one time, when certain engineers that Spidalieri felt would be needed to make the alliance successful were not included, he almost declined to be its director. However, in these negotiations, with the VP of GE and Belcan, Spidalieri, in effect, was able to plan the future organization and "write his own ticket" for the description of his position as director. The final agreement was approved for purchase by Anderson on June 15, 1987.

PROFILE OF J. J. SUAREZ

J. J. Suarez is not only the President of Belcan Engineering, but also a man with many hats. Born in Cuba, he obtained a Ph.D. and often thinks like an academic. Although trained as an engineer, he soon became an experienced manager with strong tendencies of being an entrepreneur. J. J. serves as the chief airplane pilot of Belcan and annually demonstrates his skills as Belcan's "pig roaster." With his hard work and breadth of reading he could be called Belcan's "renaissance man."

J.J. Suarez of Belcan's Airforce (cartoon by Tom Philpot)

Dr. J. J. Suarez was born in Havana, Cuba, in 1949. His father, Jesus and mother, Monina (Rodriguez) owned a general store but soon after J. J.'s birth, they entered the import-export business dealing mainly with textiles. His father had a nickname, Baraton, meaning bargaining, low price. The family moved a number of times while in Cuba, first living in a small town named Florida, a town of about 2,000 people. When J. J. was about 10 years old, the family moved to Havana. He attended Spanish speaking schools in which he learned English at about the first grade level, giving him the start to continue to improve so that his writing and operations quickly were in English.

Making use of financial contacts in New York City in 1960, J. J.'s father was able to arrange business trips in such a way as to get his family out of Cuba during the Cuban Revolution. In 1960, the family was given political asylum in the United States. J. J. went to school in Miami where teacher and students spoke no Spanish. Within six months, J. J. was writing all of his papers in English. The family moved to San Juan, Puerto Rico when J. J. was in the seventh grade. Since the move was in the fall, J. J. lost nine months of schooling. However, he was given a chance to be in seventh grade with his regular age and succeeded in catching up to the others.

After the move to Puerto Rico, J. J. worked at night and weekends with his father and mother in building the textile business, fashioned similarly to Avon distribution. After J. J.'s father cut the large rolls of imported cloth into four yard bundles which J. J.'s mother wrapped nicely and priced, J. J. used a VW van to distribute groups of 20 bundles door to door to the poor sections throughout the island. The family hired a driver of the van because J. J. was not old enough to drive at the start. Later, as the business thrived, they built up a fleet of 12 vans. After J. J. left for the U.S., J. J.'s father became wholesaler for a number of businesses which had started to compete using the same concept. Finally, the business changed to a wholesale importer of textiles for retailing.

After J. J. graduated from a university high school, he obtained a B.S. in Civil Engineering from the University of Puerto Rico graduating Cum Laude in 1971. Several of the faculty had degrees from Cornell which provided the basis later for J. J.'s obtaining a doctorate at Cornell. After obtaining his B.S. degree, he worked for several years as design engineer in the firm of Lebron, Sanfiorenzo & Fuentes, Architects-Engineers, San Juan, Puerto Rico in order to save enough to accept an assistantship at Cornell's school of Engineering. He became a naturalized American citizen in 1970.

In 1973 Suarez obtained the Masters of Civil Engineering degree from Cornell after securing a teaching and research assistantship. Later he completed the requirements for his Ph.D. in Engineering at Cornell in 1976. J. J.'s Masters thesis and Doctoral Dissertation focused on the specifications for the structural integrity of reinforced concrete structures in the event of earthquakes, especially in the case of nuclear power plants.

Soon after J. J.'s arrival at Belcan, Belcan was faced with criticisms from visitors regarding its auditorium because a column obstructed the view of the speaker in front of the room. J. J. applied his structural knowledge to the problem: he removed the load-bearing column, constructing a roof bridge between the other load bearing columns. The project had been viewed with skepticism by other Belcan engineers but the change resulted in exactly what J. J. had predicted, and the "one less" column auditorium remained to demonstrate "J. J.'s column."

In November, 1988, following the reorganization of Belcan, Suarez became President of Belcan Engineering Group, Inc., one of the two operating companies of Belcan Corporation.

In the period of 1992–1993 he shared his experience in Total Quality Leadership by delivering a presentation to the Construction Industry's Presidents' Forum regarding the overview of the Construction Industry Institute (C.I.I.) concerning state of the art Total Quality Management implementation in the construction industry. Dr. Suarez is a member of C.I.I.'s Executive Committee. He was also the Keynote Speaker at the Construction Industry Cooperative Alliance (C.I.C.A.) forum on implementing TQM in engineering and construction. He was a featured speaker at the Michigan Construction Users' Council (MCUC) membership meeting in May of 1993 and at the European Construction Industry's TQM Conference in Birmingham, England in July of 1993. He spoke at the GOAL/QPC Convention in November of 1993.

A licensed pilot, J. J. enjoys flying, playing the Bongo, and is also an avid golfer.

INTEGRATED STRATEGIC PLAN

"I continually learn from others and from life's experiences."
RALPH G. ANDERSON

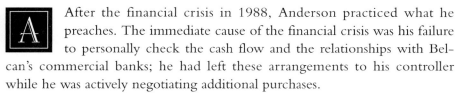

After the financial crisis in 1988, Anderson practiced what he preaches. The immediate cause of the financial crisis was his failure to personally check the cash flow and the relationships with Belcan's commercial banks; he had left these arrangements to his controller while he was actively negotiating additional purchases.

The long run causes of the crisis went beyond the short term credit problems. The rapid expansion of the mid 1980s required the expenditure of capital for doubling space on Anderson Way. Management consultants helped the trend toward diversification. Belcan entered new activities without an integrated plan.

While the basic core of the company had been selling both temporary services and in-house engineering services, the company spread out into construction of buildings and the manufacturing of products. Rapid growth of the mid 1980s moved the small engineering firm from the size suitable for a single owner's personal supervision to a middle-sized firm with sales of over $100,000,000 requiring greater attention to organization and long run planning.

Anderson moved to meet the short run crisis in 1988. He sold off operations that did not fit; he reorganized with a new management team; and in 1989 he began to develop a long run integrated strategic plan. With the help of existing and newly recruited professional management, he began to concentrate on the management of the going concern; he remained, at heart, an entrepreneur, but he showed that he could adjust to modern approaches. In fact, as an entrepreneur he became creative not only in engineering and risk taking, but he also was creative in management techniques.

In this chapter we describe Belcan's operations in the 1990s and explain the strategic planning for guiding the firm throughout the decade of the 1990s. Then we look at the present concept of lead team and agile enterprise that replaced the confusing organization structure of 1988. Next we describe in detail the concept of partnering relationships. Finally, we re-emphasize the company's central focus on Total Quality Leadership.

Corporate Structure

After the diversification of the 1980s and the difficulties of "having too many balls in the air," Belcan returned to the traditional areas of strength. All operations were incorporated into Belcan Corporation, a holding company with two operating groups: Belcan Services Group and Belcan Engineering Group, Inc.

Belcan Services Group

Specialization on supplying temporary technical services to industrial firms was the simplified, basic strategy from 1958. The new corporation, forming a part of the reorganization, simply grouped all the personnel and files into a separate building to carry on a dependable source of income and growth.

In the early 1990s, this corporation employed 1500 engineers and technical personnel who were contracted out to clients' facilities throughout the United States. The personnel file had grown from about 100 resumés in the 1960s to 200,000 resumés in the 1990s. The method of filing had changed from individual personnel folders to a comprehensive computer data bank of applicants.

The top management of the services group consisted of a team of four managers: CEO Michael McCaw, President Lane A. Donnelly, Vice Presidents John Kuprionis and Cleve Campbell. This team consisted of two professional engineers, one business manager, and one lawyer. However, each of the four top managers operated as general managers, that is, each of the four made binding decisions for the company without checking with the other three. This technique reduced the levels of the organization and added flexibility. The technique is particularly suitable for technical services since all four travel extensively; thus, one of the four could stay in Cincinnati and make timely final decisions. Of course, all four do meet periodically, as an operating board to ensure coordination and consistency of decisions and to make use of the professional specialization of each.

The operating level of the technical service organization is simple: one department of four to six people focused on recruiting new persons for the applicants' files. A second department of four to six people focused on sales and servicing the clients. The work space for the entire corporation is about 10,000 feet occupied by 30 to 40 people, filing cabinets, and computer terminals. Telephone communications are essential since, at times, the entire process of recruiting and placement is completed without a Belcan representative ever physically meeting the recruit.

The functions performed by Belcan Services Group include:

- Recruitment of engineers specifically suitable for clients.
- Sales and placement of engineers into operations of the client and follow-up services to clients.
- Routine office operations such as payment to engineers, tax withholding, insurance, hospitalization, holidays, vacations, etc.
- Billing clients for services of the contract engineers.

Regional offices are located throughout the United States. Twenty-one locations have been used; however, the offices are opened and closed as the demand dictates. Figure 7.1 shows the location of regional offices at the end of 1993.

During 1993, for instance, offices in two locations were opened and one was closed. The cost of opening and closing was relatively minor for Belcan, enabling it to adjust quickly to changing patterns. Flexibility was key to success.

The entire job of TechServices consists of dealing with people. It is one method for client companies to handle the functions of a personnel department through contracting out that function. The technical engineering job is performed within the plants of the client with the recruiting, placement, payroll performed by Belcan. This point is made to clarify the possible confusion caused by "technical services company." The personnel are technical but the services are routine functions usually carried on by a Department of Human Resources.

Belcan Business Temporaries

From its beginning in 1958, Belcan Technical Services focused on supplying engineers on a temporary, contract basis to clients. As this service expanded in the 1980s, clients requested temporary employees other than engineers. As a result, Belcan Services Group expanded its activities into supplying non-engineering personnel on a contract basis.

Figure 7.1 Location of Regional Offices, Belcan Technical Services Group

In 1992, the part of the service dealing with non-engineers was spun off and a separate company was formed: Belcan Business Temporaries reporting to McCaw, CEO Belcan Services Group. Cleve Campbell became president of this separate corporation that at first remained in the same building with technical services but later moved to a separate building. Other offices were opened: one in Raleigh, N.C., one in the Star Bank Building in downtown Cincinnati, and one in Fairfield, Ohio.

In 1993, Belcan Temporaries purchased Tri-Temps, Inc., a similar firm supplying office and light industrial workers from two locations in the Cincinnati area and one in Florida. The purchase almost doubled Belcan Temporaries in size to approximately 1,000 workers making it the largest lo-cally based supplier of temporary services in Cincinnati. Pat Moore, owner of

Tri-Temps became an area manager for Belcan. The industry was projected to grow by 14 percent but Belcan was projecting growth of 20 percent–25 percent. Social changes boosted this growth. It was felt that "employers are not looking for lifelong employees; employees are not looking for lifelong jobs."

Belcan Engineering Group, Inc.

We now turn to the part of Belcan that handles projects within the building owned by Belcan on Anderson Way and at facilities in Cleveland and Dayton, Ohio, Pittsburgh, Ashland, Kentucky, and Hato Rey, Puerto Rico. It is the part that is called "in-house" engineering and since the contracts for projects shift liability from the client to Belcan, the corporation is called a full service company. Employees of this type of company are permanent employees of Belcan but the company at times uses contract employees obtained through the service company.

We have seen that in the reorganization in October of 1989, Dr. J. J. Suarez was named President of Belcan Engineering Services, to direct all services other than contract engineers. He directs all partnering contracts; all staff departments were placed in the engineering group except for the controller, who was located in the parent company.

At this point in describing a typical formal organization, the author at first included several hierarchical charts but eliminated them because the typical bureaucratic charts would be misleading; Suarez wanted clarified, cooperative relationships with no boundaries of "teams," "business units," "centers," "projects," "clients." In short, any boundary line whether within Belcan or between Belcan and clients was out. The movement was to what we might call spontaneous cooperation and what later Suarez referred to as a "seamless" organization—both among Belcan employees and between client personnel. The most significant observation is that Belcan is typically referred to as a "high tech" company, but by the 1990s it not only led in typical engineering terms, it moved into the forefront nationally with its homogenized organization concepts of unfettered cooperation and leadership in emphasis on quality.

The Cincinnati Operations of Belcan Engineering is in two parts: process & facilities, and industrial products. Process and Facilities handles "in-house" engineering projects of great breadth and variety approaching the limits of all industrial engineering including technology development, instrumentation and controls, architectural design and facilities engineering, manufacturing systems, packaging, and material handling, construction management, cost engineering, environmental services including wastewater systems, air pollution, hazardous waste. In general, Belcan continually adds specialists to keep pace with rapid technological developments.

Industrial Products designs discrete equipment for industry such as electric motors, material handling, and robots, and provides technical writing of manuals for clients. Jim Young heads Industrial Products in addition to heading the General Electric Gas Turbine Alliance by working with a joint committee of Belcan and GE.

In 1990 Belcan obtained the services of a highly qualified specialist, Vice President Robert J. Przybylowicz to head the Human Resources group who serves both the technical services and engineering companies. (See *Profile: Robert Przybylowicz*).

By 1990 quality improvements had become a major focus nationally. We have seen in Chapter Two that Anderson has been an early leader in the ideas which became known as Total Quality Management, but by this date, Belcan led in the creation of Continuous Process Improvement (CIP) and later with the broader approach of Total Quality Leadership.

Dennis Evans developed the training program in TQL and spent most of his time holding seminars and teaching quality as a part of the management training. Evans was successful in this special emphasis primarily because the top management had been completely sold on the idea. Anderson, in the early 1980s, stressed the basic idea and Suarez as president of the Engineering Group became internationally known in the field.

After Belcan developed the CIP and TQL programs for its own personnel, it found many interested outside clients who wished to buy the program. For example, in 1993 Suarez made a major presentation in London, England and Evans taught TQL in Singapore. TQL became more than an overhead cost to Belcan, as Ralph Anderson would say, it became a "billable" service.

In 1990 Suarez prepared an integrated strategic plan for the decade built on an analogy using the Bald Eagle, the trademark of Belcan. The general framework of the plan included:

- Driving forces are trends in the marketplace keeping the eagle soaring.
- Vision aims at the ideal state or the star on which the eagle sets its eyesight.
- Mission describes the short term strategies and scope of services and markets served.
- Objectives, the basis for setting plans are the wings and feathers of the eagle.

- Values guide daily behavior: the heart of the eagle dictates conduct.
- Organization structure: defines roles and responsibilities of units.

More specifically, the details of the plan reiterates the reason that clients seek Belcan.

1. Clients can refocus resources on research, manufacturing, and marketing.
2. Belcan needs flexibility to handle peak loads while reducing cost of permanent engineering staffs.
3. Cross-pollination of technology from other industrial processes can best be insured from the outside (e.g., from Belcan).
4. Costs can be reduced by taking advantage of outside competitive rates.
5. Contracts can provide turnkey projects that are ready to produce returns to clients.

The strategic plan for the 1990s specifically mentioned the trend to enter long term strategic alliances or partnering, as opposed to contract work on a project-by-project basis. The plan envisioned that partnering arrangements will be the norm for the future in industry.

Whereas Anderson had responded to the market on an opportunistic basis using his gut feeling and "wandering around management," Suarez introduced management concepts typically used by larger firms. Anderson and his entrepreneurial style (including his continual attempt to motivate employees through daily messages on the electronic mail) adjusted to Suarez's more sophisticated management. On the other hand, Suarez seems to have consciously tried to avoid formal organization concepts in an effort to develop his own version of flexibility. The movement by Anderson to deeper attention to management concepts while retaining his tried and true entrepreneurial concepts and the movement by Suarez to stress the same flexibility for larger firms made the two styles (Anderson's and Suarez's) quite compatible.

The strategic plan was revised annually each year, and by the end of 1993 President J. J. Suarez had developed the organizational concept that was referred to as "Agile Enterprise." (See Figure 7.2) By this date, the typical box type of organizational charts were maintained in the computer memory and printed only on request for informational purposes.

The responsibility of developing, updating, communicating and ensuring compliance to the strategic plan rests with the Engineering Group Lead Team. The Lead Team is comprised of the CEO of Belcan Corporation, The President of the Belcan Engineering Group, Engineering Vice Presidents, and Directors reporting directly to the President. Each member of the

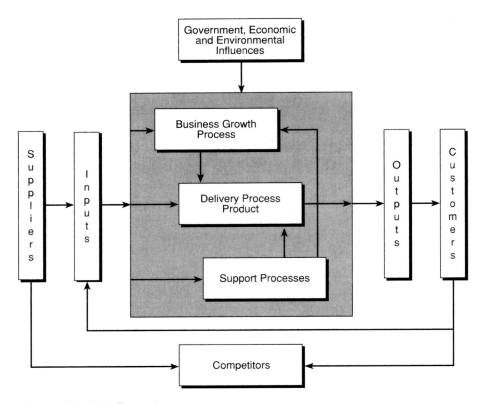

Figure 7.2 Agile Enterprise

Lead Team is responsible for ensuring that the contents of the Strategic Plan are thoroughly communicated to everyone in his or her organization. President J. J. Suarez chairs the Lead Team.

The purpose of the lead team, as described in a SIPOC* chart designed by Dr. Suarez, is the following:

TO:

- provide long term direction through strategic planning
- establish yearly business goals and plans
- monitor execution of business plans
- lead improvement efforts aligned with fundamental objectives
- allocate key resources

*SIPOC, one of the Continuous Improvement Tools utilized at Belcan, is an acronym standing for **S**uppliers, **I**nputs, **P**rocess **C**ore **F**unction, **O**utputs, and **C**ustomers.

IN A WAY THAT:

- promotes communications
- expedites decision making
- builds teamwork
- provides cohesive direction
- empowers Business Units
- sets the example for leadership
- fosters missionary behavior in Total Quality Leadership

SO THAT:

- The company can sucessfully drive toward its vision and business goals.

This team met every Monday at 7:00 A.M. with those members geographically removed from Blue Ash participating through the use of conference phone hookups. With Suarez serving as chief operating officer, in effect, operating decisions were made jointly by the fifteen members of the "lead team." The attempt was made by Suarez to recognize the flexibility gained in the past by Anderson's management approach by adapting the large organization with new flexible concepts.

In mid-1993, Suarez defined "Agile Enterprise" in the company newsletter, *By-Lines.* We quote at some length from this pronouncement by Suarez.

"Belcan Engineering Group, Inc. consists of entities designated as Business Units and Service centers. These entities develop yearly operating plans consistent with the strategic plan. Business units maintain an industry, technology, or single-client focus. Partnering relationships provide the basis for a single-client focus. Industry-focused business units provide services to a defined industry segment having a commonality of manufacturing and industrial processes, product development needs, or technology requirements. Technology business units will provide speciality services in coordination with the industry-focused business units. Service centers supply generic services to internal customers, as well as selected outside customers.

"Decentralization and a flat organization will always be pursued in order to maintain a closer link to our external customers. Our organization will be fluid. It will be periodically evaluated and readjusted to satisfy external and internal requirements. Cross-functional teams will interweave with our formal organization in order to facilitate and accelerate CIP development. Cross-functional team tasks will also be in alignment with our Fundamental Objectives. The Engineering Lead Team will assign champions to lead corporate-wide improvement efforts."

"Within this organizational structure, we must develop an " agile enterprise" in order to respond to ever-changing business needs. An agile enterprise is characterized as being:

- lean, defect-free, quick to react to new opportunities, and ready to assimilate new technologies
- able to leverage individuals and management capability to its fullest
- capable of making decisions at the right level, minimizing the time for request to go up and down traditional organizational ladders
- effective in assimilating feedback for continuous improvement
- swift in disseminating information between customer, internal groups, and suppliers
- highly interactive, allowing physically-disbursed and organizationally-segregated personnel to work collaboratively with one another
- committed to mutual responsibility for success, within Business Units and Service Centers
- driven by mutual trust, based on the need to make intergroup cooperation a first choice in sales, project execution and continuous improvement efforts
- aligned between all units in setting and pursuing shared goals, and changing pathways or removing roadblocks when problems arise
- entrepreneurial and intrapreneurial, by forming opportunistic partnerships with other companies and between internal units in order to leverage skills and to take advantage of fleeting market windows
- focused on communication networks to expedite decision-making."

In short, Suarez was attempting to express what Anderson simply referred to as his "gut feeling." Note that Suarez referred to entrepreneurial, opportunistic, fluid, and what he later referred to as a "seamless" organization. The unique nature of Belcan's relationships with its clients demanded that its engineering services should flow smoothly without regard to bureaucratic boundaries, i.e., "seamless."

Partnering Relationships (Alliances)

The technical services group provided continuity for Belcan from 1958 as the first stage of its development. In 1976, upon moving to the Deerfield building, a second stage organized the engineering division that consisted of

Ground aircraft support unit.

the "in-house" services. The third stage began in 1985 when Belcan organized its first partnering relationship (or alliance). This more advanced stage depends on long term contracts in a partnering attitude with improved means by which the client issues requests for services as a routine. Long term contracts are usually for three to five years.

Partnering relationship (alliance) is a long term commitment between two or more companies for the purpose of achieving specific benefits for the objective of maximizing the efficiency of each participants resources. The concept must be built on trust, common goals and shared values. The Construction Industry Institute (CII) has found that the concept has the following benefits: continuous improvement, reduced costs, resource utilization, increased innovation, increased efficiency, stabilized work load, enhanced responsiveness, skill development, and improved trust and teamwork of partnering relationships. Each of the alliances are quite different since each is tailored to the needs of the individual client. One should note the variety and size of Belcan's clients.

Figure 7.3   Clients of Belcan

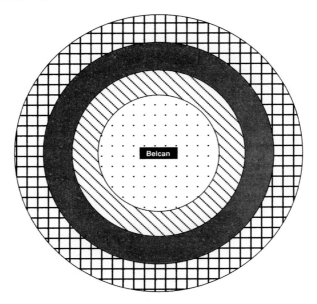

BGP (Procter & Gamble)
Eli Lilly & Company
GEAE (General Electric Aircraft Engine Division), Evendale, Ohio
GE ED&C (Electrical Distribution and Control Group)
R.E.O.P. (Regional Engineering Office, Pittsburgh), DuPont, Pittsburgh
Inland DCAT (GM Division, Dayton)
S.E.E.D. (Specialty Engineering Equipment Division), GE Lighting Division, Solon, Ohio

BELCAN ENGINEERING IN-HOUSE PROJECTS

Mead Corporation
Armco Steel Company, L.P.
Bristol–Myers Squibb
E.I. duPont de Nemours
Henny Penny Corporation
Hobart Corporation
Hobart Brothers Company
IBM
Joy Manufacturing
Pfizer Inc.
B.F. Goodrich Aerospace

GE Industrial & Power Systems
General Motors Electro–Motive Division
Martin Marietta
Nichols Institute
O'Garahess & Eisenhardt
Procter & Gamble
Quantum Chemical Corporation
Shionogi Qualicaps, Inc.
Siemens Energy & Automation, Inc.
Thomas Industries, Inc.
University of Kentucky

BELCAN TECHNICAL SERVICES (CONTRACT ENGINEERS)

Aeronca, Inc.
Allison Gas Turbine
Armco Steel Company, L.P.
Carolina Power and Light
Cincinnati Electronics Corp.
Cincinnati Milacron, Inc.
Cummins Engine
Curtiss Wright
Ethicon Endo–Surgery
Ford Motor Company
Garrett Turbine Engine Company
Conoco
Ericsson GE Mobile Communications, Inc.

Lexmark International, Inc.
Mosler, Inc.
Miles, Inc.
MCI Communication
McDonnell Douglas
Pratt & Whitney Aircraft
Rockwell International
Rolls Royce
Teledyne Precision
Texas Instruments
Walt Disney Imagineering
Williams International

BELCAN TEMPORARIES

Cincinnati Gas & Electric Company
Dyment Company
University of Cincinnati
PNC Bank
Harvest Fresh
Gibson Greeting Cards

Du Pont

In the 1980s Du Pont started to contract out its engineering design functions to four regional offices: REO-SE with 37 percent of workforce REO-H with 25 percent of workforce; REO-W with 20 percent of workforce; and REO-P (Pittsburgh) with 18 percent of workforce. The Pittsburgh office concentrated on Process and Facilities Engineering Support for Polymers produced in five Du Pont plants. Its territory extended to El Paso, Illinois, Louisville, Kentucky, Montague, Michigan, Ft. Hill, Ohio, Circleville, Deerfield, Towanda, in Pennsylvania, Washington, Potomac River, Belle, West Virginia, and Fayetteville, North Carolina.

In 1985 Belcan introduced the partnering concept into its operations when a friend of Belcan informed Anderson that Du Pont was looking for a new partner for its REO-P. In 1985, Belcan signed a five year contract to provide Process and Facilities Engineering Support. The Reo-P (Regional Engineering Office-Pittsburgh) was first managed by Dan Swanson who hired 150–200 technical people. Approximately 20 were Du Pont employees, 165 were permanent Belcan employees and 20 were contract engineers. The contract was tailored to the needs of Du Pont and when Du Pont made a change in personnel, usually Belcan had to adjust. Thus, in 1989, Swanson returned to Cincinnati in the Industrial Engineering Division. Henry Helm managed REO-P from 1989–1991 followed by Bill Ryan from 1991–1993. The Du Pont alliance served as the initial entry into long term engineering contract with large companies.

BGP Services

BGP Services was the second partnering relationship formed by Belcan after the initial Pampers contract had been completed in 1986. However, BGP was different because it had three engineering firms: Belcan, GPA (Perdikakis') and Foxx & Associates as partners for their client, Procter & Gamble. Decisions and coordination were more complicated and time consuming. For example, it took 5 months from May to November 1986 for negotiations before operations could begin. The general manager, Vince Salerno, was brought from Texas to set up an independent firm on Northland Blvd. initially hiring 300 engineers. We saw in Chapter 6 that the formation of BGP had a traumatic effect on Belcan since it took 200 engineers from Belcan's headquarters building at a critical time during which Belcan was attempting to digest the purchase of Lodge and Shipley and other efforts of diversification in the Anderson Way building that had just been doubled in size.

The financial contribution from BGP was held to a minimum by the bargaining power of Proctor & Gamble and by the fact that Belcan received only one third of the profits whereas it contributed two thirds of the engineers. During the years 1988–1990, BGP operated as a separate and distinct entity, however, because of the complexity of ownership it required more of Anderson's thought, just at a time when his attention should have been on other pressing issues.

On the other hand, BGP began to operate more efficiently. Its relations with P&G's paper division were quite satisfactory, and in 1989 the Laundry Category (soap) division resulted in expansion becoming the largest Engineering & Diesel Services contract in Greater Cincinnati for 1989 and 1990. From a low month in mid-1989 in which 256 engineers were employed it expanded to a level of more than 525. With the decentralized nature of P&G's operations, BGP actually became a double alliance, one with the Paper Division and the other with the Soap Division.

The demand for engineering design services is cyclical. For this reason, BGP's permanent employees were kept at the 300 level with contract employees being used for the expansion and peaks. This action reduced the risks of over staffing and it also should provide additional income for the Belcan group of companies since Belcan Services Group was a convenient and efficient source of temporary employees.

After Anderson bought out the shares of Perdikakis and Foxx, BGP became solely a Belcan operation with an average of 400–500 employees. BGP's building is several miles from its client, P&G, and near Belcan's headquarters. Prior to buying out Perdikakis and Foxx, operations of the alliance were completely independent of Belcan's executive; however, afterwards the general manager became vice president and a member of the lead team of Belcan's Engineering.

General Electric: Specialty Equipment Engineering Division (SEED)

In 1987 Belcan and General Electric's Lighting Division formed the Specialty Equipment Engineering Division (SEED) to service lamp-making manufacturing systems in all 25 GE Lamp Manufacturing plants worldwide.

The SEED 41,000 square foot plant in Solon, Ohio, contains an electronic and control development laboratory with 150 employees. It has a documentation vault safe containing over 1 million tracing and microfilm cards.

SEED has been awarded Automation Associate status by GE-Fanuc in its continuous motion machinery (Automation and Control Engineering). It led in Control System Technology (Computer Integrated Manufacturing) using PLC for its basic system. System integration has been developed in the field of Robots and Vision Systems. SEED maintains a facilities/assembly and testing section. The computer equipment/PC Networking uses over 100 terminals and a number of special purpose software capabilities.

The SEED operation of the General Electric alliance was managed by Ben. M. Spidalieri who had experience with GE before coming with Belcan. (See *Profile: Benjamin H. Spidalieri.*)

General Motor's Delco Products Division (DCAT)

In 1987 General Motors and its Delco Products Division chose Belcan to operate an on-site Component and System Product Development Testing Laboratory in Dayton, Ohio. Over 35 people managed by Tom Bosco provide environmental and mechanical testing to Delco Products and Inland Fisher Guide product development groups. After a successful three year contract, Belcan renewed the contract on a new, open-ended basis. Once an alliance gets through its initial adjustment between managers in two companies, the chances are good that the arrangement can continue for years.

Eli Lilly

In the summer of 1989, Belcan and Eli Lilly established a partnering relationship to provide Process Facilities engineering and design support to Lilly's Lafayette, Indiana plant. Beginning with 25 people, the partnering arrangement expanded to over 200 and then gradually reduced to less than 100 persons. In this way, Belcan satisfied the fluctuating needs of its client while keeping costs to a minimum.

Lilly, in effect, retains Belcan to perform engineering services as needs develop. This long run relationship makes it possible for both Belcan and Lilly to quickly adapt to new demands. The entire agreement is flexible and dependent upon the trust that evolves from aligning both expectations and ultimate integration of goals. Since Lilly had been one of Belcan's oldest clients, the alliance got off to a good start and continues to improve through patience, vision and commitment by all in both organizations. Upon his return to Cincinnati from Pittsburgh in 1991, Henry Helm became program director for the pharmaceutical portion of the alliance.

In 1992, Art Frohwerk developed a large portion of the alliance handling capsule-making operations. The new project, named "Boomer," involved translocation of five existing operations into two new locations, one in Spain and one in the United States. This portion of the Lilly alliance directly involved a Japanese health-care products company, Shionogi & Co., which had an 80-year relationship with Lilly. In 1966 a joint venture between the two companies, named Japan Elanco Company, Ltd., had the leading capsule-sealing and filling machinery.

General Electric's Aircraft Engine Group

In 1991 under Jim Young, Belcan's Industrial Products Engineering Division agreed to provide engineering, design, and test support for various facets of GE's business at several locations, centering in the 120 employees on the second floor of the headquarters building on Anderson Way. The turbine engine activities had been one of the first activities of the in-house projects. Thus, the alliance shifted individual projects over to a longer range commitment by each of the two companies.

Custodio, Suarez & Associates (CSA)

In May of 1992, another alliance was formed with Custodio and Associates of Puerto Rico. This alliance represented a variation from the other alliances since Custodio had itself been a firm that provided engineering, architectural, planning and construction to clients in Puerto Rico, the Caribbean, and Central America. J. J. Suarez, President of Belcan Engineering Group became an officer in Custodio with a new name, Custodio, Suarez and Associates (CSA) and later just CSA.

The CSA Belcan contract provides that Belcan Engineering will transfer technology to enhance CSA in Total Quality Leadership, Project Management and Control, and Computer Aided Design and Drafting. The alliance also accelerates the development in Puerto Rico in Process Engineering and Design Services, Automation, Control and Instrumentation Engineering, and Computer Integrated Manufacturing.

McGraw Engineering and Design Business Unit

In 1992 Belcan expanded its scope of operations when it acquired McGraw Engineering and Design, Inc. as a part of the Process and Facilities Division of Belcan Engineering Group. With its 60 employee group as a business unit reporting to Mac McCammon, Vice President of Process and Facilities with David Vice, General Manager, and Richard Logan, Director of Operations, McGraw serves the metal industry. Armco Steel Company is a major client. Offices in Ashland, Kentucky handled about two thirds of McGraw's personnel. They had completed a slab caster building for Armco handling all activities and delivering it as a turnkey contract. At the same time, McGraw had full responsibility for the construction of a slab caster for Newport Steel in Northern Kentucky. The company had expanded outside steel to aluminum facilities and used beverage can processing plant. Vice and Logan had felt that Belcan had superior skills in chemical processing and that McGraw had needed this increased strength. (See *Profile: David Vice*)

Belcan has also tried other partnering relationships. Belcan opened an office in Glendale, California to support the Walt Disney Imagineering Organization in a wide range of theme park Attraction Programs. For the first year, the contract was successful, but difficulties during the second year resulted in closing the operation. Whereas Belcan has in the past remained flexible in opening and closing offices to suit temporary needs of clients, the strategic plan for the 1990s is to build partnering relationships which last for five years or more.

To illustrate the flexible guides used by Belcan in developing partnering relationships, note some are in locations near the clients' plants (Du Pont, Lilly, & Delco); some are located in or near Belcan (GE Aircraft and BGP); and some are near a client's plant but serve plants worldwide (SEED). The terms of the contracts differ as to the manner in which Belcan is paid for the services of its employees. All personnel working in Alliances are Belcan employees.

The concept of alliances has advantages for both Belcan and the clients, but the concept is not a simple one to work out. Each client has a different corporate culture and different ways of doing things. Thus what may work in one alliance might not be suitable for another. Belcan's flexibility gives it an advantage, since its culture has historically been able to adjust to different settings. As Belcan is successful in adding more alliances, it must focus on adjusting to these other environments. Clients are usually much larger than Belcan and have difficulty in adjusting so it leaves it to Belcan to do the adjusting. However, as Belcan gets larger, if it is not attentive to the problem, it can become more inflexible.

Total Quality Leadership

In addition to the strategic plan for alliances and partnering, a second major element in Belcan's strategy has been the development of the role of total quality leadership (TQL). This element is an overall management training program for Belcan's employees and for clients. As we discussed in Chapter 2, it evolved from an earlier quality improvement program for individual projects which Belcan called continuous improvement process (CIP). The goal of the process is to change the company's culture from preconceived ideas to one of listening to what the customer wants. Training courses thus focused on customer satisfaction as that measurement of quality in a service industry.

Compatibility of Management Styles

We have seen in this chapter that Belcan in the 1990s moved to a more formal, generally accepted style of management. We have also seen that the trend toward partnering relationships requires a flexibility and cooperation that Anderson has emphasized from the beginning. His personal creed that "I can learn something from anyone" fits the needs for the future, but as Belcan increases in size, the impact on formal management approaches may cause problems. Anderson likes technical subjects but he expresses himself in an unsophisticated, down-to-earth manner. He continually emphasizes that management is 100 percent dealing with people and he prefers to deal verbally rather than through formal, legal documents. He prides himself in being able to judge a person and to trust him with "handshake" not a "signature."

Although Anderson recognized the need for the more formal organization and strategy formulation introduced by J. J. Suarez in the 1990s, he continued to supplement the structure attempting to retain the personal touch as a "wandering around" advocate. In addition, he would try out new management techniques.

In early 1991, Anderson installed a weekly "gathering" of seven of the top management for the purpose of providing an opportunity for each executive to tell about his personal life, attitudes, and aspirations. The gathering continued for several months, but was temporarily suspended because some executives became so busy that they missed the meeting and Anderson wanted 100 percent participation or no gathering.

The objective of these meetings had been the same as the objective for including the thirty-one profiles in this book. Thus, it is hoped that this

Wandering Around Management: RGA checking with Katie

book will serve as more than a business history; it has the potential of describing Belcan's culture and personalities. With this attempt the larger Belcan can longer retain some of the advantages of the smaller company.

J. J. Suarez used strategic planning and gradually implemented long range plans. Yet, Anderson had been successful in deciding for thirty years using his gut feeling. In the 1990s one could still find instances of opportunism supporting an integrated planning structure.

A general case involved opportunities in international activities. Many clients were large multinational companies and Belcan needed to adjust to the demands of their international operations. We saw that Turbine Power Systems had sold products through marketing by General Motors in the 1970s. The Pampers designing in the 1980s had international dimensions through Procter & Gamble's world wide plants. The alliance with GE in SEED after 1987 handled world wide markets in electric lamps. Lilly's alliance involved international participation with a Japanese firm and building a plant in Spain. In fact, the very nature of Belcan's services with large multinational companies necessarily involved international aspects. Thus, even without a clear target for international growth, Belcan gradually has become international.

In addition, although Anderson did not enjoy international travel, he began to check out other international opportunities independent of the thrust by multinational clients. During a short period after Toyota located a large auto plant in Kentucky, Anderson tested out possible ways that he could exploit the success of the Japanese entrance into Kentucky. This possibility quickly was dropped primarily because of the tendency of Japanese management to work with Japanese supporting firms.

A second trial in international waters involved the People's Republic of China. Through an old friend, Anderson explored opportunities in the PRC in designing in margarine production. This brief trial involved Bob Clark's trip to China as explained in Clark's profile.

A more direct trial by Anderson involved a trip to Hungary with President J. J. Suarez and Financial VP Riordan in following up on potential opportunities afforded by General Electric.

Interestingly, the personnel of Belcan had been internationalized in the recruiting of top executives who were immigrants from other countries and through executives with international experience. Without a conscious plan to become "international", Belcan had laid the foundation of people who could support international activities. Thus, Belcan further increased its flexibility to adjust to the trends of industry into international activities.

In Chapter Eight, we focus on the people who were available in the 1990s to implement not only the strategic planning developed by Suarez but the opportunistic ventures of Anderson.

We have seen that the trend toward partnering requires a flexibility and cooperation that Anderson had emphasized from the beginning. Basic to this style is that each member of Belcan's managing lead teams understand the attitudes and aspirations of each other. The experiment of the weekly gathering may not be revived for lack of time, but the profiles of the Belcan family can serve some of the needs visualized by Anderson with his experimental gatherings and this part of the book can be updated periodically to cover added personnel.

The heart of Anderson's attraction of quality personnel is in his personal attributes and a style that encourages employees to become a closely knit team through knowledge of the background of key individuals with whom they work. The idea is easier in small organizations, but Anderson stresses techniques that retain the advantages of smallness even in a rapidly growing company.

Profile of Bob Clark

Bob Clark was born on Oct. 20, 1933 in Brooksville, Bracken County, Kentucky, and attended high school where he played basketball. He received an athletic scholarship at Kansas State University where he majored in mechanical engineering. After the first year, he returned to take a job at Miami Margarine Co. as a mechanic and decided to transfer to the University of Cincinnati in their co-op program in mechanical engineering. During his last four years, Miami Margarine paid for Bob's total college program as long as he maintained a B average.

After obtaining his bachelor's degree he continued to work for Miami Margarine and began work on his master's degree. In 1966 the company built another plant at Albert Lea, Minnesota. Clark moved his family to Minnesota where he was General Manager of the plant with 20 employees; ten years later it increased to 300 employees with $100,000,000 in sales. Returning to Cincinnati in 1976 as Executive Vice President of Sales and Manufacturing, he was made president of the company in 1978. By this time the company had bought a plant in Los Angeles so the three plants provided a national network of margarine manufacture.

In January, 1991 Clark retired from the company and left the management in the hands of one of the family members whom he had trained for taking his position when he retired.

Clark and Anderson had met each other after Clark had returned from Minnesota (1976) during the planning of the Anderson Way building (1980) through a common friend, Chuck Kubicki. While Clark was active in Miami Margarine, the government of the PRC contacted him about making better use of soybeans and their oil.

In 1991 after Clark's retirement from Miami Margarine, Anderson suggested that he devote 30 hours a week to development of partnering relationships between Belcan and prospective partners in food plants and refineries. Success in this new thrust would enable Belcan to present an image: "From French fries to jet engines."

Clark talked with Anderson about the proposal to build seven plants in China (approximately $15 million per plant) whereupon Anderson sent Clark to China to check out proposals. China would generate hard currency through sales of products to India and other Asian countries that import cooking oils from the United States. The proposals were moving nicely until the Tienanmen Square episode after which nothing developed. However, Clark continued his development work for Belcan in the food industries.

PROFILE OF
WILLIAM LEVEY MCCAMMON

A After this Kentuckian had worked in North Carolina, Virginia, Kentucky, Saudi Arabia, Tennessee, New Jersey and California, he finally returned to the Midwest. In November, 1991, J. J. Suarez offered him the position of Vice President of the Process and Facilities Business Unit of the Belcan Engineering Group. Because the job would return him to his major expertise of chemical process and Division level management, Mac joined the Belcan family.

William L. McCammon, generally known as Mac, was born January 26, 1938 in Lebanon, Kentucky. His father, Donald K. McCammon, a graduate of the University of Kentucky in electrical engineering, married Evelyn Taylor, also of Lebanon. Mac knew engineering early in life through numerous moves throughout the Eastern states, including Jamestown, Kentucky during his father's work on dam sites as a power house engineer for Dravo Corporation. Later, the family moved to Louisville where his father was among the startup staff of General Electric Appliance Park where he eventually retired.

Mac graduated in 1960 with a BSCHE from the Speed School at the University of Louisville in their cooperative program in chemical engineering. On August 25, 1960 Mac married Sandra Huber from Louisville.

After graduation, Mac became a process engineer for E. I. Du Pont in Kingston, North Carolina. He gained experience in new product development, and served nine and a half years in the Firestone Synthetic Fibers plant at Hopewell, Virginia, building a pilot plant, a prototype plant and a commercial plant to produce tire cord. He holds a patent for a vessel used in the process.

Wishing to remain in the design/construction field, Mac joined C&I Girdler Company, a subsidiary of Bechtel Corporation. In 1974 he became project manager for a job with Procter & Gamble in Cincinnati with their disposables paper group (Pampers and Rely). Later with Bechtel, he became Engineering Manager of the Project Mechanical Group.

Between 1982 and 1985 McCammon served as resident manager for Bechtel's SABCO operation in Al Khobar, Saudi Arabia, performing work for Aramco Oil. Upon completion of his overseas assignment, he returned to the USA and assisted in opening up an office in Kingsport, Tennessee performing work for Tennessee Eastern until 1987. He then moved again with Bechtel to Parsippany, New Jersey to manage another start-up operation.

In order to locate in the midwest in a more stable position, he joined Lockwood Greene based in Spartanburg, South Carolina, one of the oldest engineering companies in the U.S. (in business since 1838). Mac was selected to open a new office in Blue Ash, Ohio. Between 1988 and 1991, the office grew to 75 people and was made a separate profit center. From this office, he moved the short distance to Anderson Way.

PROFILE OF ROBERT J. PRZYBYLOWICZ

 On April 9, 1990, when Przybylowicz joined Belcan as Vice President, Human Resources, he was handed a two-inch thick group of files of legal suits being defended by legal counsel. Since his arrival, Anderson observed, "Bob, we didn't have many suits against the company but now with a Director of Human Relations, we seem to have many more." Thus is the self-fulfilling prophecy of management!

Born on February 23, 1947 in Dayton, Ohio, Bob's childhood was totally in the Polish culture. He attended a Polish language school, a church with services conducted in Polish and he lived in a strictly Polish neighborhood. He had a job from an early age. His mother felt that everyone should know more than the three R's: they should know cooking, washing clothes, ironing, and sewing. His oldest sister became a nun with two master's degrees and is chief financial officer for her religious order.

Bob's father was an immigrant from Poland, coming to the mining area of Pennsylvania with his grandfather who worked in the coal mines. Bob's father did not go into the mines because he found that he could represent the company on a soccer team as his main employment. As a semi-professional soccer player, he was drafted to a German team in Dayton by a jewelry company which got him a job with General Motors.

Bob graduated from Boys' High School and entered engineering at the University of Dayton. After two years in engineering, his draft number indicated imminent call to the Vietnam War. Signing up for courses needed for deferment, he found that the only courses available were in psychology. Box decided to switch from engineering to psychology, taking two psychology courses to maintain his deferment. He accepted a job in 1969 at the Dayton Mental Health Center while taking formal classes to set up his own cooperative program. Later, he obtained the master's degree in clinical psychology at the University of Dayton. After two years with the Dayton Mental Center, he served four more years at the Eastway Community Mental Health Center, supervising counseling and community relations.

Since employment in clinics did not pay well, he decided to move into industry. In 1978 he applied to the Standard Register Company and met the usual screening techniques of a large (4,000 employee) firm. After remaining all day in the office before the personnel manager talked with him, he returned the next day for an interview with the sales manager. Later, he was given the usual personnel tests and referred to the Marketing manager. After about four days of interviewing and testing, the interviewer still said that the company did not have a job to fit his qualifications. With this delay in getting a final decision, Bob said, "If the company can't make a decision with this amount of information, I don't want to work here." The Marketing Manager responded,"Can you start next Monday?" and Przybylowicz became an analyst in its personnel and training department.

In 1980 he was offered the position of Director of Human Resources, Planning and Development for Top Value Enterprises, Inc., a company actively participating in the commercial travel industry. Later, Top Value was purchased by Baldwin United Company and time was right for another move.

In 1985 Przybylowicz became manager of Employee Services for Mead Imaging, Miamisburg, Ohio, a company in the high tech imaging industry. He managed manpower planning and training programs for top level management and high tech personnel throughout the United States. Although this later position paid well and was interesting, when the management of Belcan approached him, he was receptive. Even though the pay was not greater, he joined Belcan because of the challenges of the company and the nature of the management approaches.

PROFILE OF DAVID VICE

Born in Ashland, Kentucky, David Vice literally grew up in the steel industry. His father was manager in the Armco Steel Company in Ashland. After graduating from Paul Blazer High School, Vice won an Armco Scholarship to Virginia Polytechnic Institute from which he received the degree in Engineering Science and Mechanics in 1972. He joined Kentucky Electric Steel where he learned the practical side of steel making until 1987.

In 1988 Vice was married and worked briefly for a fabricating company. He was hired by McGraw Engineering and Construction Company, one of the largest such firms in the United States to open the Ashland office specifically to serve Armco steel.

In October of 1990 the McGraw Construction company sold the engineering and design portion to four men, each having about 25 percent ownership, Richard Logan, Sam and Gerald Mansbach and David Vice.

Richard Logan and David Vice had used contract engineers from Belcan Technical Services. They averaged about 25 engineers at a time. For this reason, Vice and Logan contacted J. J. Suarez of Belcan when they needed funds to keep operating. In May of 1992 the four owners of McGraw Engineering and Design sold to Belcan. David Vice became director of the Belcan operations handling the Ashland, Kentucky operations and Richard Logan became General Manager stationed at Blue Ash and handled Middletown, Ohio operations.

NOT ONLY THE COMPANY ONE KEEPS, BUT THE PEOPLE THE COMPANY KEEPS

"My gut feeling is my method of selecting the best people."
RALPH G. ANDERSON

With our understanding of Anderson's philosophy, the biographical summary of his life and the history of Belcan, we can now check his objectives and their achievement. A simple statement of achievement is, "we're still in business." Survival is necessary for the realization of any other objective. Profitability and growth are typically measures of business success. In concluding this story, we can quickly say that Anderson has succeeded when measured against these usual criteria.

Two objectives stated by other successful entrepreneurs frequently appear as they retire. First, "after working hard for a lifetime, we now want to enjoy retirement." Secondly, "since my success is due to certain institutions and to society, I want to return some of my profits to society."

This chapter focuses on two different objectives that are central to Anderson's thought and actions. These are not usually emphasized by other entrepreneurs. Anderson wants to have fun in his business dealings and he wants his employees to enjoy working for the company.

Anderson has stated, "Since I've been having fun throughout my working life and since I am not going to retire, I'm merely going to keep on having fun working; retirement would not be as much fun. I am now giving and will continue to give to institutions and society, not because I feel an obligation, but because giving to them is a lot of fun."

This chapter analyzes how Anderson chose the associates who are in the profiles and how these people have been attracted to Anderson. In short, how do Anderson's aspirations match with the aspirations of those who have chosen to join Belcan? The answer to this question is our chief finding because the heart of a technical services company is the location of suitable applicants and matching them with the needs of the clients. Furthermore, as Belcan grew, Anderson needed to carefully select permanent employees for his own company.

Thirty-one profiles have summarized the interviews with representative key permanent personnel of Belcan. Some generalizations will be apparent on first reading; however, others are subtle and can be best appreciated by more detailed analysis.

We will infer from the results of the selection process what Anderson was looking for. Later, we shall uncover what the employees found in Anderson's personality and style that attracted them to him and to Belcan. Anderson would agree with most of the former but he was unaware of some of the latter. In short, this chapter attempts to answer two questions. First, how did Anderson choose employees, and secondly, what factors attracted employees to choose Anderson? A successful working team is determined not only by the recruiter but also by those who are looking for an employer with whom they can feel comfortable and where they can have a sense of what Maslow calls self actualization.

Anderson's style of management emphasizes people. Belcan as a high tech company was built on drawing boards, computers, systems engineering, buildings, video, and "hard" assets but these become valuable only as people use them for service to clients. This chapter studies these people.

Anderson's Selection of Personnel by "Gut Feeling"

From the beginning of the interviews with Anderson, it became clear that he had confidence in selecting personnel with only 10–15 minute interviews, but he would explain it only by his "gut feeling." From a study of the profiles of those who had been selected by "gut feeling," the results indicate certain characteristics of what he was looking for even though he did not use an explicit checklist or questionnaire. Most of these appear more logical than the New England entrepreneur who used the "short socks" test discussed in the first chapter. Later we will see that some factors uncovered from characteristics of employees may have been thought to be important to Anderson but not stated.

The importance of what Anderson calls "gut feeling" is related to what Dr. Deepak Chopra calls "spontaneous right action." Considerable study indicates that subconscious tendencies are important in higher levels of consciousness. An attempt to rationalize factors may, in fact, be self defeating.

We shall see that many of the elements that Anderson was looking for were characteristics of Anderson himself. He was satisfied with himself and felt more at home with those who were similar to him. However, we will also see that he was not satisfied with a clone of himself; he searched for others "from whom he could learn" and would provide creativity in their relationships.

Were there subtle and underlying elements that were common to both Anderson's gut feeling and the employees that were selected? In short, did Anderson use a model of himself subconsciously in his search for people who would best fit the needs of Belcan? In what ways were the people he selected like himself?

The "Belcan Person" is a composite of elements common to Anderson and individual employees. When we take thirty-one profiles of individual employees, we can uncover some of these elements that are common to both Anderson and a representative subgroup of his selectees.

What do these people tell us about Anderson's "gut feeling"?

First, Anderson looked for people who liked to work. If their early life was on a farm where money was short, he felt that this background was evidence of a hard worker. Examples include Beymer, Gilliam, Helm, Mitchell, and Smith. Individuals who earned their own way in an engineering school indicated they liked to work. Examples include Anderson, Gilliam, Clark, McCammon, and Spidalieri, attending cooperative programs. An additional large number had other self help, e.g., scholarships, part-time jobs, etc. People who liked to start early in the morning got extra "gut" points.

A representative statement of hard work came from Bob Przbylowicz as he described the expectations he faced from a Polish family in Dayton. Work was a way of life; it was the route for an improved standard of living for an immigrant family.

Second, Anderson looked for people with new ideas. Creativity got his attention. He needed this type of person because he wanted to let the people "give it a go." Even if others would view an idea as a wild one, Ralph went for it. Anderson listened and supported such attempts using his gut feeling even if risk could not be calculated, e.g., Hope and Mendenhall.

Third, Anderson's gut feeling sought only people he could trust. If a person gave signs of ethical defects, that person was out. This quality was especially important in Anderson's case because he would give a job and let the person alone without detailed controls. But Anderson could not afford to do this if the person was not motivated to make a major effort toward the cooperative enterprise and could be trusted to work without supervision. Anderson found that the majority of those selected could be trusted; in fact, in some cases, trust was self-fulfilling: by expecting a person to warrant trust, that person will fulfill that trust.

Fourth, Anderson looked for a simple, straightforward person who is not impressed by titles or evidence of class distinction. No names or titles are on the doors to offices; only the desk name is evident. Business cards give the title for sales purposes but when asked to give the exact title, the person often had to look at the title on his business card to make sure what that title was. Anderson referred to individuals and actions, not to positions and job specifications. These people best fit Suarez's seamless organization.

In the early days of the company, the fact that a person came from Kentucky seemed to give "gut points." A graduate of the College of Engineering at the University of Kentucky was worth lots of points.

A sense of humor was not a prerequisite for employment, but since Anderson wants other people to enjoy their work, he would help them develop a sense of humor, even if it had been submerged under a business facade. He laughed easily and wanted others to laugh and enjoy life while at work. In the interviews we made no subjective quantitative scale of measuring a sense of humor but we can communicate by using a scale of one to ten: all but three or four would register at least an eight or nine. In several cases, we at first failed to notice the different types of humor but on rechecking, the humor scale increased. Apparently, Anderson believes that a happy person who is having fun is a more productive person.

The elements of "gut feeling" stated above describe most of the personnel attracted to Belcan. In addition, our further analysis uncovered classifications that describe more completely the "Belcan person."

Types of People Attracted to Anderson's Style

We can be more explicit when we interpret comments in interviews with employees as they explain the important aspirations of their lives, their interests, and their preferences. These in-depth chats indicate the types of people attracted to Anderson, not necessarily evident in the gut feeling elements discussed above.

First, we shall summarize some of the general characteristics of people in the profiles and be more specific with detailed examples of these other interesting characteristics.

Recruits from Clients

A large percentage of those recruited came from two client companies. Former employees of Procter & Gamble and General Electric Company were a core of Belcan personnel. Many of these were hired as contract engineers of Belcan but continued to be located in the clients' plants. We have seen the reason for this fact—as larger companies decided to outsource engineers, they laid off their own employees who could then more easily be hired by Belcan.

A Belcan employee who has been employed by a client became an important advantage later as Belcan began to develop alliances. The meshing of the cultures of Belcan with its clients is facilitated if the Belcan employee has been intimately in touch with the client. For example, the smooth transition of S.E.E.D. into an alliance was facilitated by the fact that Spidalieri had previously been a life long employee of General Electric. Swanson, Evans, Le Saint, and Frohwerk were all alumni of P&G.

Sports of Major Interest to Employees

Although sports had been a major interest of Anderson throughout his life, there is no evidence that he was aware of the importance of sports in the early life of up to 25 percent of those interviewed. For support of this observation recheck the profiles of Beymer, Clark, Donnelly, Evans, McKnight, Przybylowicz's father and others.

Education a Special Factor

Obviously an engineering company would be composed mostly of engineering college graduates. Several went to engineering schools with cooperative programs allowing them to work part-time in industry, as Anderson had done. But the unusual factor is that several employees were the first in their family to attend any college. See the profiles for Evans, McKnight, and Gilliam. Other employees completed their undergraduate degrees after joining Belcan. See the profiles for Smith and Wagonfield. A large number with engineering degrees returned to obtain an MBA.

Several Belcan managers have been oriented to teaching and training. Lane Donnelly had been a teacher in the East after graduating from Murray State University. Denny Evans has expanded the CIP training program for Belcan and its clients but in addition, he loves to offer university-based training sessions.

Entrepreneurs Attracted to Anderson

Anderson was not the only entrepreneur at Belcan. He enjoyed having venturesome people all around him. The diversification spree of the mid-1980s brought more entrepreneurs into Belcan. Joseph Campbell and Robert Flisik helped create Multicon and presented challenging high-tech ideas. These ideas were appealing to Anderson because they were on the cutting edge and the new companies provided spectacular growth. However, in such cases, entrepreneurs who like to operate independently are difficult to place into a harness to pull together . Some of the organizational crisis in 1988–1989 was caused by the multitude of these entrepreneurs.

One example of an entrepreneur is Anderson's son-in-law, Mike McCaw. After becoming bored with the practice of law in a typical law firm, he returned to a manager's role at Belcan during a time in which the company was expanding rapidly and in need of his expertise of accounting and law. After trying to become the #2 man at Belcan, he again became restless. He wanted to "do his own thing." Since he always loved golf, he had wanted to design and organize a new golf course—on his own. His desire to run his own show was so great as to cause him to resign from Belcan and go on his own with his own money.

The unique characteristic of Anderson's style reappears at this point. He wanted to encourage McCaw to satisfy the succession issue in Belcan but he understood that his son-in-law could not be happy unless he had been given a chance to demonstrate his own entrepreneurial aspirations. He wanted people to be happy in what they were doing even if it meant the loss of that person to his own plans for Belcan. Thus, Mike left Belcan without harsh feelings and returned later only after he had completed his own objective of building the golf course. This departure resulted in a refresher for McCaw's aspirations. His return was at a time when he could feel that he was needed and not just to return to the role of heir apparent. Upon returning, he became chief executive officer of Belcan Services Group while Ralph Anderson remained with the engineering company and CEO of Belcan, the holding company.

Two other profiles involve entrepreneurs. The one on Karl Schakel has already been identified as Andersons' role model. The other potential entrepreneur was Jack Hope. Hope's wife, Anita, dampened Hope's ideas of risk taking and so he served as the mental stimulator for Anderson even though he was not a part of the ongoing Belcan organization.

Hope's scope of interests and his quick, creative thinking ignites ideas in others even among those who know him only casually. He can out talk both Anderson and the author! The problem is that his new ideas tended to crowd out deeper development while provocative ideas surface.

The giesel has been the magnet holding Anderson and Hope in a common orbit. The initial idea was developed by Johnston, promoted by Hope, but held together via a patent by Anderson. The giesel made a long term contribution to Belcan because it was the reason that Jim Young began his long time tenure as the senior technical man at Belcan.

Both Anderson and Hope continue to have faith in the revolutionary idea of the giesel. Neither could attract the necessary $100 million development backing that it needed. Since the payoff was too far in the future, neither put "all their eggs" in the giesel basket, Hope going to Washington and Anderson to the explosive growth of Belcan in the 1980s. Thus, even long after this book is printed, the giesel, a model of which sits in Belcan's lobby, may, in fact, revolutionize engines. The nearby Belcan eagle may pick it up and fly away.

Robert Mendenhall, not included in a profile but discussed in the chronological chapters, represents another side of Anderson's entrepreneurial experiments. As an early associate of Anderson's, Mendenhall teamed up with Anderson several times to form abortive corporations. The first was the ill-conceived Aero-Math. Anderson may be too quick to assume risks on ideas posed by others, but Mendenhall tested him as to whether he could drop a bad idea. Anderson passed the test by dropping it within two weeks. Even without the help of others, Anderson found the Mo Gard to be an unsound venture.

Anderson became known as a guy who would "grub stake" you. Mendenhall tried again. He introduced the idea of portable power systems for developing countries. This time, Anderson did not interfere with his young Belcan organization but segregated his risks into Turbine Power, Inc. After agreeing to provide finances for a reasonable time and leaving sales to Mendenhall, Anderson concluded that he had put his money on a limited idea and a losing horse. This case merely indicates that Anderson's gut feeling may have been fair game for a wild idea, but he had the additional guts to call it quits before he could get hurt.

J. J. Suarez had been an entrepreneur in Puerto Rico, and was married to an entrepreneur with whom he had worked just before he came to Belcan. In 1992 he not only developed an alliance in Puerto Rico but he became a part of the client organization. J. J. had developed the concepts of management and strategy so that he was available in 1988 at a critical time to organize the diverse activities and to spin off the unprofitable extraneous parts that had been assembled. Anderson's skill at this time was to choose J. J. as president of a company that had become larger than Anderson's style could accommodate. Furthermore, throughout the discussion in previous chapters, we have noted that Anderson did not change his approach but merely allowed J. J. to construct his more formal design with its agile organization and strategic focus. It is clear that these opposite styles are compatible at the top of Belcan Engineering without the fact being apparent or important to Belcan employees. Furthermore, Anderson had been able to retain Suarez, the manager, and to satisfy his entrepreneurial tendencies.

Company Loyalty and Long Term Tenure

Two profiles represent an attraction by loyal friends to remain with Anderson: John Kuprionis and Bob Smith. Anderson is a businessman who develops long term friends within the company and stands with these friends over the long pull. Smith and Kuprionis are two of the long-tenured engineers who remained at various roles through many changes. They serve as cement for Belcan morale. They offer a refreshing example of a family-like relationship.

Importance of Family—A Special Attraction of Many to Anderson

Many others at Belcan looked to the company as related to a family atmosphere. Because Cincinnati is a large city, the families of employees did not know each other on a regular basis except for the annual company picnic. They did view Ralph Anderson as more than an employer. They took personal interest in the welfare of the company, especially during the time of crisis in 1988–1989. All felt that the company still was small enough to have interest in their welfare.

The importance of family to individuals shows up in two profiles as a special type who are attracted to Belcan. Larger organizations may take a mechanistic and bureaucratic approach; Anderson's style of organization thrives on the humanistic characteristics of team players. The stories of Cleve Campbell and Henry Helm stand out.

Cleve Campbell throughout his career adapted his employment to the health problems of his family and changed jobs to fit family needs. Just at the time that he had the opportunity to continue to advance in another company if only he move from Cincinnati, he resigned from that company and became available to an offer from Belcan, which enabled his family to remain in the Cincinnati area.

Henry Helm considered the preference of his German wife for Cincinnati with its German atmosphere in his decision to join Belcan. Helm's family was always important in his career decisions and his religious views were compatible with the positive thinking of Ralph Anderson.

In the cases of Campbell and Helm, Anderson's style of personal interest attracted a certain type of person. That type considers the broader aspects of life as opposed to a strong ambitious climber of the corporate ladder.

A third and more recent case was that of W. L. McCammon who had traveled extensively in foreign countries. Finally, he decided that Belcan suited his personal preference in order to be with his family for a greater part of the time. In this way, Belcan attracted an experienced engineer with foreign experience by appealing, at a certain stage in that person's life, to a broader interest of enjoying life.

The fact that J. J. Suarez is a manager in the United States is directly related to the success of his parents to place his interest high on their priorities. J. J. had helped the family build their own business in Puerto Rico.

Art Frohwerk* made the move from California to Cincinnati because he and his wife wanted their children to be close to their grandparents and get to know them well.

Ben Spidalieri, as an Italian immigrant moved to the U.S. with his parents. Each member of the family had to consider the overall welfare of each member of the family. The reason that Ben ended up in Cleveland with GE was directly related to moves based on family welfare by his parents.

The story of Bill McKnight is one of pursuing education because of the strong desire of his father to see his son get a college education.

Experienced Engineers and Managers Working Past Retirement

Belcan tapped a valuable source of skilled engineers that large companies tend to neglect—experienced, professional people who could retire but who wished to find a challenging position tailored to their abilities and interests. Active people who do not want to sit in the Florida sun and vegetate seek flexible organizations that can expand their useful lives in satisfying endeavors. Anderson, who himself is "beyond retirement age" provides the environment to offer challenges to such "senior citizens." In this way, Anderson

*The translation from German (Frohwerk—Happy workplace) is particularly suitable to Anderson's stress on employees being happy.

is like Edwards Deming who became a world famous quality expert only after the age of 65.

Two profiles illustrate how Anderson's continued interests in new ideas and his flexible organization make use of valuable service from people who might have retired.

K.O. Johnson brought with him his distinguished service with General Electric's Evendale plant. His professional qualification and his knowledge of GE made him a natural for helping Jim Young and the partnering arrangement with GE Johnson's move to Belcan is a perfect example of how an arrangement can be the gain for all concerned: for GE, because the large company can continue to take advantage of his experience; for Belcan because it can build a stronger alliance and strength as an engineering company; and for K.O. because he was able to move easily into an environment in which he could continue to make contributions regardless of his age.

Bob Clark was able to step from the CEO of a large food manufacturing company to Belcan's vision of expansion into his field of expertise. Ralph Anderson jumped at the opportunity to get Bob Clark, even though his field was outside that of Belcan's activities. Anderson was receptive to taking risks if the right person and opportunity came along.

Strength from Diversity

A diverse group of people were attracted to Anderson's characteristics and style. The scope of Belcan's activities is so broad that it needs a diverse group of people who can adjust to each client's culture. Although Anderson did not plan or look at quotas of different types of people, the result by the 1990s was a diverse bunch. The profiles show that Kuprionis is from Lithuania, Bob Przybylowicz is from a Polish family, Henry Helm is from Germany, J. J. Suarez is from Cuba, and Spidalieri is Italian.

Diversity is predominant in the types of engineering specialties employed. Each alliance demands not only a different mix of engineering specialties but each alliance must adjust to the organizational culture of the client firm. Spidalieri is closely suited to S.E.E.D. because he had been a lifelong employee of General Electric. Salerno was actually chosen by the P&G management. Helm had close associates who worked for Eli Lilly. Jim Young came from General Electric's gas turbine operations. Fitting the Du Pont culture became a more difficult problem and generally required changing the Belcan manager whenever a change was made in the counterpart at Du Pont. As progress is made in developing other alliances with clients having diverse cultures, Belcan will need to retain the flexibility of fitting its diverse personnel to the new situation.

The past experiences of Belcan people range over wide and diverse activities:

- From Jim Young's participation in the birth of the gas turbine (jet) engines to Jack Hope's attempt to build a nuclear aircraft engine
- From Henry Helm's work with CERN in Switzerland to the Belcan's 150 contract engineers' work for a period with the super collider in the Texas multi-billion project
- From work in Hollywood for Disney to the development of machinery for producing Pampers for Procter & Gamble
- From Bob Clark's margarine experience with opportunities in China to Belcan's work with Eli Lilly in Indiana
- From Mac McCammon's new product synthetic fiber development with Du Pont to Isaac Gilliam's development work with Mead Paper

Strength from Similarity

Some profiles illustrate that Anderson feels most at home with people who are "just like the boss." Some of those discussed in the profiles accentuate certain of Anderson's qualities and thus extend his image in the growing company. Examples include Harvey Mitchell and the Anderson Circle Farm. Interviews with Mitchell demonstrated this similarity. At times Harvey made general statements that could easily have come from Anderson. Mitchell seeks to add new techniques to farming and enjoys keeping the land attractive by maintaining black fences, barns and new products.

As one of the youngest in the profiles, Pat Wagonfield's preferences fit the Anderson model: he responds to needs as they evolve, such as making wooden and plastic models to speed in the design of machines for P&G's Pampers project; he retains his early interest in playing with a band, or producing a video for Belcan. He responds to new challenges and adapts to new situations. His span of attention is relatively short because there are many interesting things pressing for his attention.

Issac Gilliam is another Belcan specialist who mirrors Anderson's interests. Gilliam, an architect, gave special effort to remodeling Walnut Hall soon after he arrived at Belcan. Later, he worked on the building of the Spring House from timbers from an old barn. For several years, Isaac filled the role of helping Anderson in extending his attention to new endeavors. When the Anderson Way building was expanded, he was the key architect for designing the new facility.

Anderson remained a small town boy after he built a big city company. He was an interface between the small town approach of Harrodsburg, Kentucky and urban Cincinnati. When he makes his nineteenth century mansion available for community affairs, such as tours and benefit balls, he not only spends money on the projects but he spend hours of his own time in support of community activities. He enjoys implementing new projects both in the small town of his birth and in air conditioning for the cat house at the Cincinnati Zoo.

Management of Human Resources is the Whole Ball Game

In the service industry, using primarily professionals, the attraction of the "right" person is all important. As the company increased in size, Belcan remained current by hiring a vice president to focus on not only compliance with the laws but with use of the more highly sophisticated tools of human resource management. The proof of the success of these efforts can be discovered by studying the actual people as they perform their jobs. Thus, this chapter has summarized and interpreted the profiles of key people in Belcan.

The matching of Anderson's gut feeling with the personal desires and interests of those attracted to him has produced a functioning, compatible organization that can best service Belcan's clients. In fact, Belcan's advantage is that it can provide this service better than the clients can themselves.

The interesting fact is that a world class, state-of-the-art, high tech organization has been successful not merely by its technical superiority but by its superiority in dealing with people.

Even with the best people, a business firm must face the "bottom line." Revenue must exceed expenses. Anderson has faced the real world of business as an interesting challenge and therefore he has fun.

It is more fun to be making a profit than operating at a loss. In the final analysis Ralph Anderson is a good businessman; yes, he is an entrepreneur but he had proven his business qualities by remaining in existence for over 30 years. He may not be the best manager but since he plans to live to 120, he still has 50 years to become a good manager of a large firm.

Anderson focuses continually on whether an employee's activity is "billable." If there is one word that his associates hear as much as positive thinking, it is the word "billable." As firms become larger, more people merely increase overhead in the simplest terms: overhead consumes cash; billable activities provide the cash and income that is critical for a viable business firm. "Unapplied time" is the continual threat.

Throughout this book we have reported that Anderson wants to have fun himself and for others to enjoy themselves. We conclude this chapter on the people factor in Belcan by clarifying a possible impression that Anderson is a lucky, nice fellow, but a soft touch in bargaining. Those who have dealt with him first comment that he will do what he says he will do and that he is honest; nevertheless, he has always been a tough bargainer and a guy whose ego gives him confidence. Ralph Anderson is a hard-nosed businessman. When Ruth and Ralph had little extra funds, Ruth was the level-headed keeper of the cash. It was she who pointed out that they needed to put funds aside and invest them. It was this move in the late 1960s that started the farm purchasing campaign over a twenty-year period.

One of the chief advantages of a technical services company is to provide rapid adjustments to changes in the economy. It is clear that if such a company were hesitant in releasing unneeded people, the technical service company would have less advantage over the large, slow moving industrial firm. Since the very idea of a temporary or technical service company is the speed that they can adjust to both upturns and downturns in business conditions.

In the summer of 1993, like many other businesses, Belcan had to tighten its belt, that is, they had to release 25 employees. The four major executives sat down on a Monday and quickly responded to the most recent figures. In a firm of Belcan's type the speed of reaction is the critical factor. Through the years, Anderson has demonstrated that he likes people but when chips are down, he is not afraid to cut costs. Personal feelings cannot override good business. Furthermore, those who are close to Anderson know that they must be honest and follow through with what they say they will do. Once he finds that person does not warrant his trust, he acts swiftly. Furthermore, he seems to never forget and thus he encourages others to "do him right."

FIFTY-ONE PERCENT LAUGHTER, THE REST WISDOM AND LOVE

"Entrepreneurs are productive; gambling and lotteries are not."
J. L. MASSIE

The title of this final chapter was extracted from a letter written by the Reverend Dr. Robert Schuller of the Crystal Cathedral in California, after a visit to Anderson's farm in the spring of 1992. The full quote was "what a precious time I had with you! How I love your spirit and style! I'll remember you as the first person to whom I said, "Fifty-one percent of the sounds coming from your face is laughter . . . the rest is wisdom and love." This comment accurately sums up the manner in which the Andersons affect the people who know them.

The fact that this book evolved from an oral history project helps to explain its strengths and weaknesses. Over 90 percent of the information reported in the biographical material is based on audio taped interviews (i.e., oral history). Most of the profiles are based on taped interviews with the individuals. Standard vitae or resume material were used for checking dates and spelling because modern resumes seem bureaucratic and lifeless. The result is a "live" description of the Andersons, the people in Belcan, and many others who gave interesting observations about the Andersons and their company. The foci are on behavior, culture, interactions, motivations, and styles rather than channels, technical descriptions, memos, and orders.

Ralph Anderson: Likeable and Interesting, A "Now" Kind of Guy

The overriding personal factor of Ralph Anderson is the fact that he and his wife, Ruth, meet life together with a unified viewpoint. They operate as a team. Ralph and Ruth came from different states but they view life with the same perspective. They remain proud of the low economic status from which they started. They are proud that they have worked hard and earned a good life together. They continue to plan activities jointly. On weekends, they travel together and enjoy companionship even though each does his or her "own thing" during the week.

A person such as Ralph Anderson thrives on assuming risks, therefore making him a good entrepreneur. This gives him the freedom to lead his own operations. He enjoys teamwork, in sports and in business, and he prefers to develop solutions unhampered by restrictions and red tape of large organizations.

An associate describes Anderson as a "now" guy—a man who wants action immediately and who will strive to seek results now, not later. He is a man of action, and the action is fired by unique and creative ideas.

Though trained as an engineer, Anderson found the characteristics that are typical of an engineer were not his major advantage. His specialty is technology development but he prefers dealing with people more than things. He recognized early in life that *quality of service to the customer was the foundation for success.* This is not just a pleasant-sounding catch phrase, but a continual search for what the customer wants and the quality that the customer has a right to expect.

This story is about the founding and subsequent growth of a high-tech engineering company. It is the story of a number of people who came together and enjoyed the same type of challenges as the man enjoys. The chief competitive force of the company was the attractive environment nurtured by the chief executive officer.

Learning from the Birds

Throughout his life Anderson has learned from observing things around him. Mari Otto, a former secretary to Anderson, relates this episode.

"Once I checked to see if Mr. Anderson was free. Usually when he did not want to be disturbed the door would be closed. This time the door was open and he was standing quietly, looking out the window. I did not disturb him and waited for about fifteen minutes. When I returned, he was still looking out of the window. Finally, after checking several times, I became concerned and asked if everything was okay. He said, 'See those birds building the nest, look how they go about it.' " Anderson always said he could learn something from anybody but the statement had not been made broadly enough—he could even learn something from the birds!

One of Anderson's particularly appealing characteristics is his sense of humor. Anderson has been known to kid people and encourages others to kid him, keeping him "on his toes." Unlike many people, he enjoys stories which make him the brunt of the joke. The author's tendency to enjoy laughing at himself was matched by Anderson's effort to match or excel, even to the point of making himself the bigger joke.

One such incident involved the planning for a house on the Mercer farm. He explained that his architect had done an excellent job of designing the cabin using timbers from an old barn. The windows had been painted a distinctive shade of green, and a shiny varnish had been applied to the beams; futuristic-looking appliances had been selected for the kitchenette.

When Ruth and the interior designer saw the result, they disapproved of the green paint and the high sheen on the old timbers; they considered them inappropriate to the rustic plan. The paint was changed, the varnish was stripped away and some of the appliances were changed. Ruth added some personal touches which were truer to the Shaker motif. She suspended an heirloom wedding-ring quilt from the loft balcony, and filled the niches with crockery and baskets of dried flowers. She nestled a couple of her beloved dolls here and there, and placed an antique spinning wheel at the focal point of the large common room. On the deck she hung a strand of wind chimes with a deep, melodious sound. "There," she thought, "that's more like it."

In relating this incident, Ralph chuckled at himself, recalling his cardinal rule that everyone in the company should look to the customer and seek his or her wishes. "Now," he said, "I find that I don't practice what I preach. At the start of this project, Ruth and I had decided that the house was hers; she was the customer. I had talked with the architect and interior designer for their expertise, but failed to seek the wishes of the customer—my wife. It is hard for all of us to learn the simple things in life."

Leadership Style

Anderson's style of leadership can be best summarized as a "walking around" advocate, cutting through paperwork of a growing organization, and attempting to retain the "personal" touch among his managers. For example, an annual pig roast is held for all Belcan employees on the Mercer County farm with president J. J. Suarez serving as chief cook. Such activities tend to keep the flavor of a smaller firm even when it is becoming the largest of its type in the state of Ohio. Although he fully supports the formalized organizational techniques of J. J. Suarez, Anderson seeks to retain the feeling of the small firm and the personal touch.

In the 1990s Belcan moved to a more formal, generally-accepted style of management. We have also seen that the trend toward partnering relationships requires a flexibility and cooperation that Anderson has emphasized from the beginning. His personal creed that "I can learn something from anyone" fits the needs for the future, but as Belcan increases in size, the impact on formal management approaches may cause problems. He likes technical subjects but he expresses himself in an unsophisticated, down-to-earth manner. He continually emphasizes that management is 100 percent dealing with people and he prefers to deal orally, rather than through formal, legal documents. He prides himself on being able to judge a person and to trust him with a handshake, not a signature.

This restatement of Anderson's management style serves to reemphasize that it is this style that enabled him to build a company focused entirely on services of people to people, not the production of products. He is an engineer by trade, but he thinks and acts as a non-technical person. He stands out as an entrepreneur, not because of his engineering skills or patents of products, but because of his skills in human relations.

Anderson is comfortable seeking academic contacts but he continues to be non-academic in his own approach. Any strategic, long-range plan that is routine and sophisticated will not impress Anderson, the individual. Yet, when one analyzes the strategic plan of the 1990s, and the type of managers that are placed at the top of engineering services, one can only marvel at Anderson's adaptability to different styles and his perception that the company at its larger size needed a style that was different from the one he used when the company was smaller.

We have seen that Anderson enjoys trying new technology. He is technically minded but he does not pursue technical sophistication when it is not practical and will not pay off. He observes that too many engineers have sought technical achievement for its own sake. He jokes about his avoiding the skills to use a personal computer, repair an auto or fix something in the house. He encourages others to excel in these technical matters but he prefers to keep his perspective. For example, in the 1980s, CAD (computer aided design) took over the designing chores of the company. Anderson quickly moved to the advanced CAD's and replaced the old drawing boards. When a young computer expert suggested selling all the old drawing boards, Anderson declined. He reasoned that at times he may need the old technology and he wanted to have them available as backups. He feels the same way about automatic gadgets. He feels it is wise to retain a back up for most technological advances.

Anderson varies from the prototype of the hard driving executive who has all the answers and expects others to fit into his thinking. He is easy to talk with and not intimidating. He continues to listen to anyone who has an idea and gives them a chance to try out that idea. Repeatedly, subordinates agree that he approves their ideas but adds "one should be sure of the probable success because Anderson expects results, not theories."

Anderson Differed from Other Wealthy People

Many of the possessions acquired by Anderson in the last ten years might mislead the reader to jump to conclusions, such as that he is a "showoff," that he wants to "compete with the Joneses," that he wants publicity, or that he seeks status. None of these observations is true or, at least, none were overriding in his efforts to acquire money.

With the purchase of a Rolls Royce, a Southern mansion, and all manner of technical "toys," a casual observer would tend to classify these in what Thorstein Veblen called "conspicuous consumption." While one might accurately conclude that the evidence is clear, further study indicates that his motives have always been deeper, even though they might appear to be ostentatious.

Anderson likes to improve things. The extra money spent on black fences, interior decoration for a large house, renovation of an historic building, and other improvements are based on his desire for perfection. When close friends are asked, "is Ralph a showoff?" the answer is a resounding "No!" In fact, they often act as though that was the first time that they had ever thought of such a question. Anderson likes technical gadgets and loves owning the unusual. He likes unusual clocks and unusual mechanical or electronic things.

As we stated earlier, at seventy years of age, Anderson was reliving the portion of his boyhood when he had no toys to play with because he had to work. Anderson never wants to retire. He wants to continue to play with his toys: Belcan, his 3000 acre "sand pile" in Mercer County, historic buildings, and his constant search for new developments in engineering and archeology. Others may play monopoly as a game; Anderson plays monopoly with actual property. Others play with play money, Anderson plays with real money. Anderson has always enjoyed watching sports, and even with this, he is not passive but active in his support. More recently he complained that his grandson had beaten him in racing go-carts around the circle in front of Walnut Hall.

Many of his daily actions indicate that he does not visit spots just to be seen. For example, he has a favorite spot for lunch which prepares food to fit his diet; although he belongs to "clubs" he visits them only when he entertains a person who *does* like them. He contributes to his favorite charities and to educational institutions. Though his name might be publicized, he is more interested in the fact that the charity received his assistance.

Anderson's contributions in the 1990s increased in size and number. Although even with a small income, he gave selectively to his favorite charities. The favorite recipients were to his alma mater, the College of Engineering at the University of Kentucky, his home town community of Harrodsburg and Mercer County, community projects in Cincinnati, and to the ministries of Robert Shuller's Crystal Cathedral.

The largest two sums were $2,000,000 to the capital fund of the College of Engineering, University of Kentucky and $1,000,000 to a chair at the International School of Christian Communication, Crystal Cathedral Ministries, California. Anderson enjoys responding to needs as he observes them personally to such varied uses as professional planning of the campus of the University of Kentucky, an alumni promotional film for the College of Engineering, a public park in Harrodsburg, books and research papers for the hometown library, air conditioning for the cat house at the Cincinnati Zoo, Boy Scouts, and computers, sorely-needed items for a Harrodsburg Hospital, and a significant contribution to the Shakertown Fund. Anderson has fun making money but he also has fun giving it away!

Pathological Narcissism? No!

A psychologist of Harvard Medical School, Steven Berglas, has made a study of "pathological narcissism"—a disease suffered by certain successful people. Such people could include Donald Trump, Pete Rose, Gary Hart, Imelda Marcos, Jimmy Swaggart, Jim Bakker, Ivan Boesky, and Michael Milliken. Typically, these people have a tendency toward self-destruction. Berglas observed that wealthy people who have this disorder usually exhibit four qualities: arrogance, a sense of aloneness, the need to seek adventure, and adultery. This classification is presented as a means of showing that Ralph Anderson does not fall into the category of some wealthy people. *He is interesting because he is unique.*

Anderson was never an arrogant man and he continues to learn what he doesn't know (he jokes about his difficulty with spelling words). Anderson likes to be around people and to work with people; he likes to take risks that have a chance of paying off but he never would gamble just for the adventure of it. He has never bought a lottery ticket. Ruth and Ralph came through life together with no thought of any "playing around."

In Chapter 6 we saw that Belcan and Anderson received publicity on the business pages of newspapers in the 1980s just prior to the major financial crisis, but in the 1990s Anderson repeatedly received awards for his lifetime work. In 1992 he was named Alumnus of the Year by the College of Engineering, University of Kentucky and in 1993 he was installed in the College's Hall of Distinction, which placed him as one of ten from among 15,000 graduates to receive this honor. Furthermore, Anderson was given the Harrod Company Award by the Chamber of Commerce of Harrodsburg, Kentucky for his long time efforts to improve his hometown community. This latter award was based not only on financial contributions to the hospital, library, theatre, and park, but for the renovation of three houses on the National Registry and Anderson Circle's effect of beautification of the country. Finally, the honorory degree from the University of Kentucky topped all.

Belcan: From the Past . . . and Into the Future

Belcan's continuity, then and now, depends upon an entrepreneur who searches for opportunities and takes risks that benefit himself *and* society. The challenges of the 1990s can best be understood with a summary of how the company reached this last decade of the 20th century.

Throughout the evolution of Belcan's short history, its success depended upon trying many approaches and quickly dropping the ones that appeared to be dead-end ventures. Some dead ends were creative, some were high-tech. Several were based on patents held by Anderson; most were high risk ventures. The names of some of these risks include Mo-Gard, Turbine Power Systems, Aero-Math, Giesel, and Ulticon.

The profitable ventures, generally speaking, were not high-tech ventures. Profit were obtained from routine activities such as supplying engineers to large companies on a temporary basis, testing the products designed by others, providing full service engineering to clients more cheaply than they could do it themselves, and crash programs in designing production machines to fit the clients' tight marketing targets.

Why Has Belcan Been Successful?

Anderson, along with most successful entrepreneurs, possesses *a strong desire to work for oneself.* Anderson worked in three large companies. He succeeded during the 1950s to fit into these hierarchies, but he felt that he would have greater satisfaction and would move faster if he could act freely on his own initiative. During the years of working for others, he made a number of contacts that later paid off after he formed Belcan.

He established a respected *reputation* as a project engineer with several leading industrial firms; his superiors liked him, his reputation for making a fair deal made others desire to work with him, and his readiness to take risks when others shied away led him to having his own company. In 1958, even though he had no personal capital, he incorporated Belcan.

The discovery of a niche was a key to the creation of new opportunities. This niche was not a new product nor a patent on a new process. It was the recognition that large manufacturing companies required, at times, to contract out engineering services. In Ketco and Allstates, he found smaller companies that had demonstrated there was a niche for independent engineering firms. He observed how Karl Schakel and others had made a success of their own firms. He had entrepreneurial mentors and prototypes to study.

Anderson's *distinctive competence* as an engineer was a prerequisite for success. Anderson believes himself to be a good salesman, yet he believes that engineers in general are poor salespersons. He feels they tend to focus on technical matters rather than human skills. Anderson's past experience includes more than technical skills; people like to work with him. His distinctive competence is his orientation toward the customer. Human factors were more important than technical matters.

Finally, *a service that others will pay for* is essential for a profitable business. After World War II, large corporations wanted to hire engineers on a temporary basis since their projects and contracts required only temporary service. Anderson had developed skills in selecting engineers and placing them successfully. The result would be a company that contracted engineers on a temporary basis to other corporations for their projects performed in the customers' own facilities. If Belcan could perform this service more cheaply than the customer could, the company would be profitable. Chances of success were improved by the following advantages:

1. It did not require much start-up capital—which Anderson did not have.
2. It served at the very beginning as a means of ensuring Anderson an avenue to obtain work for himself.

3. The derived advantage was that Belcan could staff its own company with permanent employees who had been tried out on a temporary basis with other companies.

In 1976, Anderson observed a structural change in industry. In the 1960s and 1970s, large companies found out-sourcing to be preferable to expensive permanent engineering staffs. Industry would contract with Belcan for projects to be performed in Belcan's buildings. The in-house capabilities built by the mid-seventies made it possible for Belcan to satisfy the needs of large firms that were thinking of reducing their fixed costs of a permanent engineering staff.

The expansion of Belcan in 1976 was the acceptance of engineering contracts on a project basis for the customer in Belcan's facilities. This change made Belcan an engineering service company in addition to a service company for supplying engineers. The major effect of this change was that Belcan required additional capital for operations. With this change would come increased risks but also increased profits. The problem with this expansion was that Belcan would need to concentrate on selling its services to cover increased fixed costs. A business slowdown for clients would be felt more directly by Belcan since Belcan would have the fixed costs with the loss of flexibility previously maintained by the smaller company.

In 1985, a new opportunity appeared in industrial organization nationally. Some large companies wished to disband their entire engineering departments in certain parts of the company and to make long term contracts with engineering speciality companies for all projects within the defined limits. This change moved Belcan from the field of temporary engineers of projects to one of long term cooperation between Belcan and its customers.

Such a relationship became a joint operation in which Belcan cooperated with its customers in improving quality and efficiency. The idea that a smaller, specialized firm could join with a larger one through "partnering" or alliances evolved as co-operation among companies took different forms. This trend, started by the Japanese in the 1960s, appears to have only just begun and thus when we look to the year 2000 and beyond, broadening the concept of alliances should continue, not only for Belcan, but for all of American industry.

We have seen how this third stage of strategies began with Du Pont and later increased to more than five alliances. These opportunities continued to grow by development of alliances with more companies but also by more alliances with the same customers.

Management Succession

Private ownership by a single person or family is always faced by the need to plan for management succession. Typically, this problem is not faced because entrepreneurs do not usually look into the future for fifty to one hundred years. Also, the effects of estate and inheritance taxes can be the dissolution of the company as governments demand payments in cash at the time of death of the owner.

Generally, one alternative solution to the limited life of an owner is to "go public." This alternative involves making the company desirable to investors on the stock market. The advantages of a public company are (1) that at the death of the owner, the inheritance and state taxes can be paid from cash obtained by selling stock on the market, whereas if the company remained private, it may need to be liquidated or sold. (2) As a company succeeds, it continually must have more equity capital since borrowing has its limits. (3) Operating executives can more easily be given a part of the action.

In the case of Belcan during the 1984–1989 diversification spree, consultants and professional advisers suggested reducing risks by expanding into other activities. The management of Belcan learned its lesson of straying from tried and proven specialized activities. If going public requires diversification, it may be harder to convince Belcan's management that it is worth it.

A second alternative would be to sell the company. Similar technical service companies would be interested; however, often a firm that has been built by one person may not be worth as much if that person was not a part of the sale. Belcan, in fact, is buying other engineering companies.

The most likely alternative for Belcan is its continuation as an independent firm. Although at this writing, Anderson has passed the retirement age of 65, he has not even considered retiring. In fact, he has stated often, only half jokingly, that he doesn't plan to die or that he plans to live to 120 years. Anderson will probably continue to be CEO of Belcan for as long as he is in good health.

The family succession is now well determined. Candace is the only daughter of the Andersons and Michael McCaw, her husband, is already CEO of Belcan Services Group and could easily take over as CEO of Belcan, the holding company. With his legal and accounting academic background and more than 10 years experience working with engineers, McCaw would become the manager of Belcan's existing strategies, even if he operated in a more conservative manner, taking fewer risks than Anderson has in the past. Suarez most likely will acquire greater equity as he already has in CSA, the Puerto Rican operation.

Table 9–1

EVOLUTION OF BELCAN'S MAJOR STRATEGIES, 1958–PRESENT		
Engineering Specialist for Industrial Firms		
1958–Present	Technical Services	Temporary Engineers—1–3 years
1974–Present	Engineering In-House Services	Specific Projects—1–5 years
1985–Present	Partnering (Alliances)	5 year contracts (open-ended)

Financial Analysis—Belcan and Ralph Anderson

Since the focus of this study is on a privately held company, it has avoided the discussion of the "bottom line." We outline the growth of Belcan in Gross Revenue but we have not provided annual profit data (See Figure 1–1). A major advantage of a privately held company is that profit data is none of the business of the general public and the author never did even ask Anderson for this information. He was so open and trusting in his relationships that the author felt a moral obligation to protect the interests of such a person. The long run financial picture is quite simple. Anderson started the company with little savings and barely enough to live on. Within thirty years he clearly made a lot of money. The formula was simple: he worked hard, saved as much as he could each year, began investing in farm land when the price was much lower, and used as much borrowed money as he could. He convinced bankers that he would be a good risk in operating Belcan and if one bank turned him down, he succeeded in finding another bank. The federal farm credit bank was an obvious choice for financing farms. He clearly believed in leverage. At times he had trouble with lenders because he obviously enjoyed taking risks with his own money and others tended to be more conservative.

Several financial facts are generally known:

1. Farm land in central Kentucky has increased in value over the thirty year period. On the average, a person buying blue grass land thirty years ago could be expected to realize considerable capital appreciation.

2. Belcan provides engineering services that could be called "producer goods" by economists. Producer goods face a business cycle because its demand is a demand derived from consumer goods. Thus, without

checking any accounting records of Belcan, an economist would expect to find some years in which the firm would show little or no profits. This has, in fact, been the case with Belcan.

3. Anderson has observed that he is faced with a three year cycle that he has tried to smooth out without much success. In the final analysis, entrepreneurs such as Anderson do not have a faint heart and in cases of cycles tend to buy low and sell high rather than the general tendency of buying when things are good and selling when things are bad. Often entrepreneurs operate by the "contrarian rule" as we have seen repeatedly in Belcan's case. That is, if times are bad, it merely provides a good time to invest more and to work harder.

Entrepreneurship

The purpose of this book has been to describe how one man succeeded in creating a company by developing a concept that was compatible with the industrial environment. The description moves from the individuality of Ralph Anderson to the growth of the company, Belcan. We have noticed that Anderson used an early employer as his role model of how he could exploit the concept. We have described his genealogical background and the pioneering spirit that surrounded him without implying that these were determinants; they are merely facts to be interpreted.

Entrepreneurship is not a professional endeavor. Anyone can try to start a business, but only a minority succeed. How can any book help a person start a new business? It is obvious that the first step is that the person must desire strongly to go into business for himself and work hard to make the business succeed. A book can only confirm that desire in the reader and provide a role model for the achievement.

Anderson the entrepreneur possesses several characteristics that are prerequisites for success:

1. He is competent in his speciality.
2. His personality encourages people to work with him and to trust him.
3. He works hard and has pressed for success.
4. He not only didn't mind taking risks, he actually sought productive risks.

Other characteristics reinforced these necessary ones:

1. He has fun in working out answers to problems. It is a game for him.
2. His wife supports him completely in what he does.

colorplate 1 Ralph G. Anderson, Entrepreneur. (Painter, David Mueller; Photographer, Beth Gibson)

colorplate 2 The Experimental Giesel and Patent Holder.

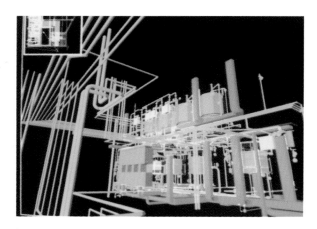

colorplate 3 Computer-Designed (PASCE) Plant by Belcan Engineering.

colorplate 4 Belcan Lobby Headquarters with Ohio, Kentucky, and United States Flags.   (Photographer, Beth Gibson)

colorplate 5 Michael and Candace Anderson McCaw and Family—Matthew, Jason, and Amanda. (Photographer, Beth Gibson)

colorplate 6 President J. J. Suarez Cooking at the Pig Roast.

colorplate 7 Painting of Walnut Hall, Anderson Circle Farm. (Painter, James Werline; Photographer, Beth Gibson)

colorplate 8 Wildwood, Home of Harvey Mitchell, Anderson Circle Farm. (Photographer, Beth Gibson)

colorplate 9 Anderson Circle Farm in the Winter, 1993. (Painter, Edith Clifton; Photographer, Beth Gibson)

3. He is not only self-confident but he feels that even bad results carry with them the silver lining for success.
4. His intelligence is average to good but his attitude is that he can learn from others and from his experiences.

Overall Summary and Conclusion

To summarize the essence of this story about Belcan, Anderson, and entrepreneurship, we will look at the several necessary factors for success. The following three may not be sufficient but they are absolutely necessary:

Anderson makes lifelong friends whom he can depend upon and who can depend on him.

First, *trust and personal contact are essential*. Technical achievement may help but it is not as important as the human factor.

Second, *optimism and positive thinking* provide confidence in oneself and result in self-fulfilling prophecy. When things go wrong, as they usually do, optimism reduces stress while increasing effectiveness and efficiency of human actions. Luck plays a large part in all activity, but with intelligence and optimism, a person creates his or her own luck by locating where opportunities turn out to be lucky.

Third, *one must face risks*. Risks are pervasive in life. Those who enjoy taking risks have the potential of improving their own situation and are basic to the improvements in society. We have described how Anderson jumped at taking risks when others who might have been as well qualified technically preferred to play it safe. The author found few of Anderson's friends who were envious. In retrospect, many people believe they could have taken the same risks and could very well have become as successful as Anderson. They recognize, however, that the key is in taking that risk.

EPILOGUE

Entrepreneurship is PRODUCTIVE gambling! Gambling and governmental lotteries are unproductive. The difference: too many risks already exist in society and nature—someone must assume the risks (insurance companies, governments, entrepreneurs). Gambling involves the creation of a risk that did not exist before the lottery, or the parimutuels were added to racing, or the game cards were added to athletic events. Entrepreneurs usually prefer to take risks that are potentially productive; gamblers get hooked on creating risks so that they can take them. Entrepreneurs produce food, transportation, buildings, and clothing; gambling, per se, produces ONLY risks. An economist would say that the marginal value of the dollar wagered is more than the marginal value of the dollar that *might* be won.

In industrialized countries, such as the U.S., we are fortunate to have productive risk takers such as Ralph Anderson. Underdeveloped countries and fatalistic societies have a crying need for people who are willing to take productive risks. All seem to have plenty of people who will buy a lottery ticket or gamble in an unproductive manner. The ultimate secret to increasing the general welfare of human beings of the world is to seek out, encourage, and reward the productive risk takers while discouraging unproductive ones such as gambling.

Maybe an answer to many of the world's economic problems simply involves showing people who like to gamble that they can enjoy and gain from taking productive risks. In short, society might convert unproductive gamblers of the world into productive entrepreneurs! One step toward this objective is to have more books such as this one to identify and describe role models for productive risk takers and fewer books on the success stories of gamblers.

ANDERSON CIRCLE FARM

A Anderson bought his first farm in Mercer County on October 24, 1967. Known by its former owner's name as the Bissett Farm, it consisted of 220 acres on Curry Road. Both Ruth and Ralph wanted a program for saving in the long run; they reasoned that farm land in Ralph's home county was a good place to invest. Anderson had always loved the county in which he was born. Anderson's sister, Gladys Dean, still lived in Harrodsburg and her son was already a farmer there. During the first years, Bobby Stratton, Anderson's nephew, managed the Anderson farms. During the next 12 years they added three other farms (Ballard, Venters, and Parson) consisting of 541 acres contiguous with one another on Morton Road which was several miles from the Bissett Farm.

A farm adjacent to the Bissett Farm consisting of 136 acres, known as the Bonta Farm, was purchased on July 20, 1978. Later, a smaller farm (Balden) was bought on October 17, 1980. The Bissett, Balden, and Bonta farms were better for his use than the three on Morton Road. From October 1980 to March 1985, Anderson suspended his farm buying program to the ones already purchased and operated by the family.

In March 1985, the purchase of two additional farms, Kennedy and Gregory (Walnut Hall), set the stage for the integrated farm operations on a professional basis. With the Walnut Hall farm, located at the intersection of Warwick and Mundy's Landing Road, came a house built in 1830 that could be renovated. It was at this time that the name Anderson Circle Farm was selected and a black fence was used for identification.

Anderson bought the Patterson farm (with the Wildwood house) in 1985. After the restoration of Walnut Hall, the Andersons started to use it on weekends and for special parties.

But the purchase of farms did not cease. The Spilman and Tobin farms were added in 1987; four farms (Burgin, Cinnamon, two from Wilkinson) were bought in 1988; the James and Cecil farms were added in 1990; and one (Keller) was added in 1991.

Restoration of Walnut Hall

Soon after the purchase of Gregory's farm, Anderson began the restoration of the house built by David W. Thompson in 1830. The financial resources made possible by Belcan's success made it possible to carry out a program of expensive renovating of the old house. Isaac Gilliam, a chief architect at Belcan, drew up plans and supervised the remodeling. Later, Anderson sought one of the outstanding interior decorators of Cincinnati, David Millet, to furnish the house with authentic furniture, draperies, carvings and pictures.

At the time of purchase, Walnut Hall had no electricity, air conditioning, bathrooms or kitchen. The renovated house actually consisted of two separate houses that were joined. All woodwork was made of walnut from trees grown on the farm. The woodwork had been painted; three months were required to strip and refinish it. Floors were made of poplar. Ceilings were 14 feet high. There are 12 fireplaces in the house, 10 of which have been placed into working order. The heating and air conditioning were both located in the basement (which had to be dug out since it had low ceilings). A large upstairs porch became a favorite spot for the Andersons since the view from its windows overlooked the country for miles around.

During this period, Anderson became interested in his family genealogy and in the historical heritage of the county. Several Indian graveyards had been known to be on the property. Kentucky was the hunting grounds for the Shawnee Indians. Two cypress trees near Shawnee Springs are said to have been used in religious ceremonies. Interest in renovation and history of the area became a major interest of the Andersons.

As acquisitions of individual farms increased, Anderson developed a plan for the ultimate integrated farm. In general, he would not purchase land unless it fitted this plan. On the other hand, he would not aggressively seek a particular piece of land; he merely would have his real estate agent make it known that he was interested. If someone was interested in selling, he left it to them to seek him out.

In looking back over the twenty years, Anderson observed that the cost of purchasing additional farms required relatively small amounts of cash. As another farm was purchased, he would mortgage the total land at the newly appraised value. It was the increase in appraised value that contributed to purchases of much of the land.

Visits to the Mercer County Farm became Anderson's weekend schedule. The Andersons would leave Cincinnati on Friday (often Thursday evening) and return early Sunday morning. On these visits, Ralph would make decisions about the farm operations and both would use the country home for a place of rest and relaxation.

Wildwood

The expansion of the Anderson Circle Farm was a long term plan. After the purchase of Wildwood in 1985, Anderson allowed Mrs. Patterson to remain in the house several years—a house that had been in the family for more than fifty years.

Wildwood was built in 1857. Its history had received considerable attention because of a book written by Frances Jewell McVey and Robert Jewell, *Uncle Will of Wildwood,* published by the University of Kentucky Press in 1974. Frances McVey was a descendant of Will Goddard and the wife of Frank L. McVey, a president of the University of Kentucky.

The Jewells enjoyed describing the unique characteristics of their uncle. One that stood out was that Uncle Will was always in a hurry. Driving a horse drawn buckboard, Uncle Will drove around corners in Harrodsburg so fast that those walking would have to look carefully or they would be hit. Uncle Will set the pattern for the new owner of Anderson Circle who was also known as a fast driver on Interstate 75. He noted the spots where state police would check and slow down when he passed those spots. Uncle Will would have been proud of Ralph Anderson.

The Wildwood House was remodeled for the family of General Manager Harvey Mitchell (See *Profile: Harvey Mitchell*). The farm office was located at the rear and side of the house. Other houses located on purchased farms were either renovated or torn down.

During the 1970s and 1980s other purchasers of land were active in the area. Wallace Wilkinson had developed a large farm across Highway 127 before he became governor. One piece which he had acquired on Anderson's side of 127 fitted into Anderson's long range plan, thus he purchased it in 1988. Earlier, Bunker Hunt had purchased land in adjoining counties. The fact that Anderson was a native of Mercer County enabled him to deal with "hometown" people. In fact, his real estate agent and several Mercer County authorities had been classmates of Anderson's.

With his new acquisitions of land, Anderson joined a number of the "old Kentucky families" and although he continues to live in Ohio, he is not considered to be an absentee land owner (he was on the farm every weekend), and he thought of himself as Kentuckian.

Anderson Circle Farm

The employment of Harvey Mitchell in 1987 marked the maturing of the farm as a profit–making operation. As is often the case, farms are held by persons with high incomes from other activities as a means of making tax deductible expenses and thus reducing income taxes. The rules of the IRS require that the operations be profitable over a certain number of years to avoid the hobby status. The tax implications of the farm as a personal asset of Anderson, along with Belcan, of course, were important, but he had no idea of operating anything at a loss. It is interesting that at times, however, when the operations of Belcan were disappointing, Anderson would comment to Belcan executives that if Belcan did not shape up, Anderson Circle Farm would be more profitable. Mitchell noted, however, it should be made clear, "we try to make a profit all of the time but it is hard. Considering Anderson's requirement of keeping the appearance at a peak and the farm environmentally acceptable, it was very hard to make a profit consistently."

The farm emerged as a privately operated farm that attempted to do things on the cutting edge of modern farming methods. The personalities of Harvey Mitchell and Ralph Anderson resulted in continual trials of new farm methods and equipment. For example, a tobacco barn was constructed that could accommodate two batches of drying tobacco by adding a large fan to speed the drying process. Four installations of silos and continuous mechanical handling of feed can be seen on a quick tour. The manure is collected, liquified and handled through a continuous process of spreading the compost on fields. This advanced system enables Mitchell to avoid depending on chemical fertilizers while recycling organic wastes. In short, the objective is to have a completely organic farm production.

Farm Operations

Using 1993 as an example to give an idea of a continually changing farm operation, we can summarize some facts. In spite of a blizzard, a hailstorm, a flood, and a thirty–day drought, operations were quite satisfactory. By

1993 Anderson Circle Farm consisted of 25–28 former farms, 55 miles of fences, 5 miles of hard surface roads, 12 houses, 60 buildings and 7 professional staff members. They include Greg Robey, assistant manager and herdsman; Keith Driscoll, crop man; Clovis Roy, equipment operator; Mike Scull, cattle feeder; Stewart Moss, pure bred cattle employee; and Danny Rawlings, utility operator.

Anderson Circle fences were painted black, which immediately delineated the land owned by Anderson. As soon as the sale was closed on a new purchase, the black fencing was constructed. A cost analysis indicated that black fences were cheaper to maintain than white fences; however, a functional wire fence could serve at about 50 percent of the cost of the black fence. Incidentally, a green fence or any color other than black is very expensive; this note is added because a horse farm neighboring Anderson Circle Farms is owned by a northern paint industrialist who uses green fencing.

In 1993, the farm produced the following:

Crops

- 500 acres of corn
- 300 acres of soybeans (for seed)
- 250 acres of alfalfa
- 82 acres of tobacco (leased)
- 1400 acres of permanent pasture
- Hay is one of the most profitable products—"square bale the best, roll the rest."
- 400 acres of wheat (straw)

Livestock

- 1 Bull (10% ownership)
- 100 pure bred Angus cows (including premium ones, such as Diamond Dutchess with values exceeding $10,000).
- 4,000 calves and feeder cattle for beef (the number varies seasonally).

Anderson and Mitchell have evolved clear, long term policies on farm operations. Anderson immediately focuses on improving the appearance of the ground and buildings. The farming is as clearly organic as at the time attainable; Harvey states that he is trying to keep "chemistry" away, i.e., the chemical fertilizers and weed killers, and other environmental problems. Modern "no plow" farming is used except in the leased tobacco. The labor intensive growing is leased where the subcontractor must hire large amounts of labor. Harvey comments he wants nothing to do with immigrant labor; however, he cannot do anything about the type of labor that the subcontractor uses.

During the summer, from four to six college students are hired on the farm as laborers. No labor is picked up in town, as is often available for smaller farms. Bill Driscoll repairs and constructs pipes and plumbing on a contract basis. Jerry Wilson, of Harrodsburg, remodeled all houses on the farm, including Wildwood and the new spring house.

Anderson Circle (Show) Farm

After its completion, the Andersons used Walnut Hall for a number of large parties. Entertaining, thereafter, centered on the farm in place of the house in Indian Hill, Cincinnati.

A Christmas party included guests from the neighboring localities and other distinguished guests. This party was held in the house with unusual Christmas decorations.

A yearly party is held each fall for the employees of Belcan in the form of a "pig roast" with executives of Belcan serving as cooks. The master chef is the President of Belcan, J. J. Suarez. A number of the "lead team" (officers of Belcan) use the party as a means of demonstrating an old Cuban custom and recipe for a pig roast. The cooking process occupied twenty–four hours before the party and was itself the focus of a small party of six cooks who became apprentices of the old skills. In such a pig roast, the cooking is itself a party. A custom–built roaster, built by employees of Belcan, was held solely for the fall occasion. The "pig roast" is one chance for the big city families to go to the country and small town and to build up the cooperative basis for a "seamless" organization.

This company party of 400 to 600 persons usually started at noon on a Saturday with buses and cars arriving from Cincinnati. In the early afternoon, there are a series of hay rides and guided tours of the key parts of the farm. Small groups were shown the house by Ruth. A band started to play for dancing in a metal storage barn prior to supper. A large tent served as the dining room. Bonfires were built when the sun went down and served as an ending to an old time party of the type held on these farms in the nineteenth century.

At times, international visitors to the Bluegrass were shown key Blue Grass horse farms, racetracks, and other attractions. Walnut Hall was added to

Inside Anderson's reconstructed house, near Shawnee Springs.

this list of show farms to serve as an "operating farm." Although it could not be said that it was a *typical* operating farm, it is true that it certainly was operating.

Ralph and Ruth were always receptive to the use of the farm for charity purposes. Several small projects were held in the first years; however, in 1992, large gala balls were sponsored by the *Harrodsburg Herald*, thirty local couples and the Andersons with formal dress for the benefit of the Ragged Edge Theatre, the local theater for plays and musicals. A large tent was erected to handle the guests who attended. Each year 300–400 people attended paying $50 to $100 per individual or couple. Dancing, cocktails, and hors d'oeuvres were provided at the black tie event.

The Spilman Farm and Shawnee Springs

With the purchase of the Spilman farm, Anderson bought a most historic spot. Shawnee Springs had always been the center of Indian activity in the area. Near the springs was the location of one of the first cabins and houses in the state. Although the villages of the Shawnee Indians were north of the

Ralph G. Anderson observing digging of Ray's 1780 house.

Ohio River and Shawnees were nomadic, they considered Kentucky to be theirs for hunting. Shawnee Springs, as the source of Shawnee Run, was identified by the first settlers in the area.

In 1991 Anderson constructed a cabin on the farm on a hill near Shawnee Springs and a cave. The cabin was clearly to be for Ruth and her decorating decisions. The "new cabin" was constructed with beams from a barn which had been torn down using the wooden pegs that had previously been used. Plans were prepared by Isaac Gilliam; interior decoration was supervised by David Millet. Shaker type of furniture and period furniture collected from throughout the country was used in the interior. A collection of family quilts were displayed from the loft.

The improvements and further developments of Anderson Circle Farm were revised periodically as Anderson broadened his interests in the surrounding countryside. For example, the Harrodsburg Historical Society and particularly Anderson's old friend, Frances Board Keightley, had collected interesting facts about the area. Anderson donated the authoritative "Draper Papers" to the Harrodsburg Library with Frances being the continual user.

As this book goes to press the fact that the Spilman farm was purchased by someone like Ralph Anderson is becoming more significant. Renewed interest in the early history of Mercer County is centered around Shawnee Springs and Anderson Circle Farm. For example, David Williams Station has been generally located behind Walnut Hall. The hill by the lower springs clearly was the spot of an 18th century house.

McGary came to Kentucky from North Carolina in 1775. In 1778, McGary and his two stepsons named Ray erected a stockaded station at their settlement on "Shawnee Run at its Headwaters in Shawnee Springs." Excavations will determine the details of where houses by Hugh McGary and James Ray stood.

Anderson's hideaway in 1991 was located on the same hill that McGary had chosen for "McGary's Station." McGary arrived in Kentucky along with James Harrod and Daniel Boone. Being an impetuous individual, he acquired a reputation of conflict among the settlers and is generally criticized for his role in the settlers' Battle at Blue Lick, August 19, 1782, when 60 of 182 Kentuckians were killed with 50 percent coming from Harrodstown(burg).

During the construction of the spring house in 1991, Anderson extended his interests to uncovering early history. As a true entrepreneur, it is hard to determine just where his interests will lead him. The next stage includes the opening of the cave nearby. Anderson never rests in starting projects that interest him.

PROFILE OF HARVEY MITCHELL

 Ralph Anderson recruited Harvey Mitchell in the early spring of 1987. With only a phone call, a six hour interview in Cincinnati, and a day on the farm, Anderson and Mitchell signed a contract. Anderson depended on recommendations from Dean Charles Barnhart of the College of Agriculture, University of Kentucky and his "gut feeling" during chats with Mitchell.

The personality, philosophy and style of operations of the two men are similar. Both strive for success but are not hindered by adversity. Both speak their own minds; both make decisions clearly and promptly; both enjoy working together to improve the appearance of the land and to try new techniques.

Mitchell was primarily hired as General Manager of the Mercer County farm operations; however, his role became much broader. He served as advisor to Anderson in the purchasing of other farms, the renovating of buildings, constructing and maintenance of fences and roads, and "showing" the farm to visitors. While the farm generally operates at a profit, Mitchell supports Anderson's goal of improving appearances to Mercer County farm land and in experimenting with new methods and crops. Both Mitchell and Anderson thrive on risk taking.

Mitchell was born on a farm near Princeton, Kentucky, in Caldwell County in September 1950. By the time of Harvey's birth, his father, Millard Mitchell, acquired a 300 acre general farm after starting as a tenant farmer. Each of Harvey's grandparents grew up on farms. Harvey and one cousin are the only ones of a dozen grandchildren who have remained in farming.

Mitchell had not planned to go to college but enrolled in the Community College in Hopkinsville, completing his course work in three semesters. Without funds to attend the University at Lexington, he contacted the director of the Princeton farm, the University of Kentucky Experimental Station, who found a job for Mitchell working with horses on the Maine Chance Farm. In January of 1970, Mitchell moved to Lexington and majored in Agricultural Economics at the University of Kentucky.

Mitchell was married during his senior year. He accepted the position of Assistant Trust Officer for the Ohio Valley National Bank in Henderson, Kentucky. In this position he supervised the assets of 19,000 acres of crops and cattle. In June of 1973, at the age of 22, Mitchell became superintendent of the West Kentucky Research and Extension Center of the University of Kentucky where he administered state and federal funds for two years.

By 1975, Mitchell became partner in a farming operation of 1,400 acres in Boyle County where he sought to gain ownership of his own farming operations. Differences with his partners caused him to leave Danville and in 1979 he acquired the Crittenden Seed Company in Marion, Kentucky, where he farmed 80 acres of his own, 950 acres of leased land, and cleaned seed. Two years of drought and increasing interest rates during this period made profitable operations impossible; the farm was sold in June of 1983.

In March of 1983, Mitchell left self-ownership and became co-manager of Dobbs Seed and Grain Co., Inc. of Hardinsburg, Kentucky. The fertilizer and chemical portion of the company was sold to the large agribusiness company, Cargill. It was from this position that Mitchell returned to Central Kentucky and the Anderson farm in the spring of 1987.

Mitchell's wife, Lucinda, was employed for several years at the Haggan Memorial Hospital in Harrodsburg but later moved to the Humana Hospital in Lexington. The Mitchells have two children, a son, Matthew, born November 1975, and a daughter, Rebecca, born in March 1977. They live in the renovated Wildwood. Mitchell has some acreage in Casey County which he keeps for wildlife refuge; he and Matthew camp four to six times a year on his acreage. In 1992, with some help from Anderson, Mitchell purchased a farm in Mercer County.

Harvey is an active professional at the national level of the cattle industry. One area involves Quality Assurance in the cattle industry. He is away from Anderson Circle several times a year giving talks at the national level. Matthew accompanies his father to several of the national meetings.

HIGHLIGHTS IN THE LIFE OF
RALPH G. ANDERSON, ENTREPRENEUR

July 19, 1923	Born in Harrodsburg, Kentucky
May, 1941	Graduated from Harrodsburg High School
September 1941–March 1943	Employee of Curtiss Wright Aircraft, Cincinnati
March 1943–November 1945	U.S. Air Force; to Flight Engineer, B-29
1946–1947	Cooperative Program, Engineering University of Cincinnati
March 27, 1948	Married Ruth M. Tucker of Middletown, Ohio in Lexington, Kentucky
August, 1950	Received BSME from College of Engineering, University of Kentucky
August–November 1950	Barber-Coleman Corporation, Rockford, IL.
November 1950–March 1952	Test Engineer, General Motors Corp., Dayton, Ohio
March 1952–March 1953	Test Engineer, General Electric (Evendale)
March 1953–October 1958	Project Engineer, Kett and Ketco, owned by Karl Schakel
March 1, 1956	Daughter, Candace, born
October, 1958	Founded Belcan Corporation, Cincinnati, Ohio
March 11, 1960	First patent, Mo–Gard
March 1963–November 1964	Project Manager, Allstates Design and Development, Inc.

June 6, 1966–April 1968	Opened Belcan offices in Indianapolis for design of TF-41 Engine for Allison Division General Motors
1967	Purchased first farm in Mercer County, Kentucky
1970	Obtained patent: Giesel and government development contracts
1970–1977	Turbine Power Systems, Inc.—Portable turbine generators
1974	Delivered Gas Turbine Generating Sets to Egypt and Iraq
April 1976	Initiated in-house engineering in Deerfield building
January 1980	Moved into building on Anderson Way
1984–1986	Pampers project for Procter & Gamble
1985	Renovated Walnut Hall, Anderson Circle Farm
1985	Purchased Multicon, Inc. from owners Campbell, Flisik
1985	First partnering—Du Pont—R.E.O.P. (Regional Engineering Office, Pittsburgh) in Pittsburgh, Pennsylvania
1986	Formed BGP, three partner alliance with Procter & Gamble
1987	Purchased Lodge & Shipley
1987	Alliance with GE—S.E.E.D. (Specialty Engineering Division) in Solon, Ohio
1987	Partnering with General Motors, DCAT (Dayton Center for Automotive Testing) Dayton, Ohio
1987	Formed Ulticon, Inc. with J. Yankoff
Summer, 1988	Cash crunch
Fall, 1988	Reorganization: Belcan Corporation, Holding Company with operating groups: Belcan Services Group and Belcan Engineering Group, Inc.
1989	Established partnering relationship with Eli Lilly
1992	Incorporated Belcan Business Temporaries

May 1992	Acquired McGraw Engineering & Design, Inc.—Middletown, Ohio and Ashland, Kentucky
May 1992	Formed strategic alliance with Custodio, Suarez & Associates (CSA), Puerto Rico
1992	Donated $1,000,000 to International School of Christian Communication, CA.
April 1993	Elected to the Hall of Distinction, College of Engineering, University of Kentucky
1993	Donated $2,000,000 to College of Engineering, University of Kentucky
May 1994	Awarded Honorary Doctor of Engineering by the University of Kentucky

APPENDIX

C

LISTING OF ANDERSON'S FAVORITE AUDIO TAPES AND BOOKS

Audio Tapes

1. *The Excellence Challenge* by Tom Peters
2. *Even Further Up the Organization* by Robert Townsend
3. *Sell Your Way to the Top* by Zig Ziglar
4. *The Higher Self: The Magic of Inner and Outer Fulfillment* by Deepak Chopra, M.D.
5. *What Do You Really Want for Your Children?* by Dr. Wayne Dyer
6. *What Every Young Person Should Know* by Earl Nightingale
7. *The Universe Within You* by Dr. Wayne Dyer
8. *The Power of Perseverance* by Dr. Mike Wickett
9. *The Power of Positive Thinking* by Dr. Norman Vincent Peale
10. *The Strangest Secret* by Earl Nightingale
11. *Life Hope and Healing* by Bernie S. Sieger, M.D.
12. *How Winners Do It* by Michael Merger
13. *Positive Attitude Training* by Michael Broder, Ph.D.
14. *The Science of Self Discipline* by Kerry L. Johnson
15. *Personal Excellence* by Kenneth Blanchard
16. *How To Gain Power and Influence with People* by Tony Alesandra
17. *The Journey to Enlightenment* by Shakti Gawain
18. All of the Nightingale–Conant Insight Tapes

Books

1. Blanchard, Kenneth, *The One Minute Manager Meets the Monkey,* New York: Morrow, 1989, (137 pages)
2. Bristol, Claude M., *The Magic of Believing,* New York: Pocket Books, 1948, (180 pages)
3. Chopra, Deepak, M.D., *Perfect Health,* New York: Harmony Books, 1990 (327 pages)
4. Chopra, Deepak, M.D., *Creating Health,* Boston: Houghton Mifflin, 1991, (224 pages)
5. Dyer, Wayne W., M.D., *Real Magic,* New York: HarperCollins Publishers, 1992, (270 pages)
6. MacLaine, Shirley, *Going Within: A Guide for Inner Transformation,* New York: Bantam Books, 1989 (263 pages)
7. Pascale, Richard Tanner, *Managing on the Edge,* New York: Simon & Schuster, 1980 (350 pages)
8. Schuller, Robert H., Ph.D., *Life's Not Fair, But God Is Good,* Nashville: T. Nelson, 1991, (272 pages)
9. Schuller, Robert H., Ph.D., *Tough Times Never Last, But Tough People Do,* Nashville: T. Nelson, 1991, (237 pages)
10. Schuller, Robert H., Ph.D., *The Be Happy Attitude,* Waco, Texas: Word Books, 1985, (231 pages)
11. Schuller, Robert H., Ph.D., *Believe in the God Who Believes in You,* Nashville: T. Nelson, 1989, (297 pages)
12. Schuller, Robert H., Ph.D., *Success is Never Ending, Failure is Never Final,* Nashville: T. Nelson, 1988 (244 pages)
13. Schuller, Robert H., Ph.D., *Living Positively One Day at a Time,* Old Tappan, N.J.: F. H. Revell Co., 1980, (394 pages)
14. Schuller, Robert H., Ph.D., *It's Possible,* Old Tappan, N.J.: F. H. Revell Co., 1978, (160 pages)
15. Schuller, Robert H., Ph.D., *You Can Become the Person You Want to Be,* New York: Hawthorne Books, 1973, (180 pages)
16. Walton, Sam, *Made in America,* New York: Doubleday, 1992 (269 pages)

INDEX